TRIGONOMETRIC IDENTITIES

TRIGONOMETRIC IDENTITIES

by

Chhaya Pandey | Urvashi Paswan

Wordpen Academics
Publishing Science That Matter

ISBN: 978-93-6084-300-7

Published by : **WORDPEN ACADEMICS**
4736/23, Ansari Road, Darya Ganj,
New Delhi -110 002
Phone: +91-011-35004336
E-mail: info@wordpenacademics.com

Printed at **:** **Replika Press Pvt. Ltd.**

Exclusive Distributor: ***Bio-Green Books, New Delhi***

Preface

Trigonometric identities serve as the foundation for understanding the relationships between angles and sides in trigonometry, providing essential tools for solving equations, simplifying expressions, and proving mathematical theorems. From the Pythagorean identity to the double angle and half-angle identities, these fundamental relationships form the basis of trigonometric calculations and applications in fields such as physics, engineering, and mathematics. Mastery of trigonometric identities not only enhances problem-solving skills but also deepens one's understanding of the underlying principles of trigonometry.

This book, titled "Trigonometric Identities," is a comprehensive guide to understanding and applying the fundamental identities of trigonometry. Drawing upon the expertise of mathematicians, educators, and researchers, this volume offers insights, explanations, and practical examples for mastering trigonometric identities.

From the basic trigonometric ratios and their properties to the complex relationships between trigonometric functions, "Trigonometric Identities" covers a wide range of topics essential for building a solid foundation in trigonometry. Through clear explanations, step-by-step derivations, and illustrative examples, readers will gain a deep understanding of the principles governing trigonometric identities and their applications in solving mathematical problems.

Moreover, this book explores the historical development and cultural significance of trigonometric identities, tracing their origins from ancient civilizations to their modern-day applications in science, engineering, and technology. By examining the diverse contexts in which trigonometric identities are used, we gain a greater appreciation for their utility and relevance in various fields of study.

As we embark on this journey into the world of trigonometric identities, it is our hope that "Trigonometric Identities" will serve as a valuable resource for students, educators, and anyone seeking to enhance their mathematical skills. Whether you are a high school student learning the basics of trigonometry, a college student preparing for advanced mathematics courses, or a professional seeking to refresh your knowledge of trigonometric identities, this book invites you to explore the beauty and utility of trigonometry and unlock new insights into the world of mathematics.

Chhaya Pandey | Urvashi Paswan

Contents

1

Introduction

Trigonometric identities are the equalities involving trigonometric functions and hold true for every value of the variables involved, such that both sides of the equality are defined. In this mini-lesson, we will explore trigonometric identities. There are three primary trigonometric ratios sin, cos, and tan. The three other trigonometric ratios sec, cosec, and cot in trigonometry are the reciprocals of sin, cos, and tan respectively. How are these trigonometric ratios (sin, cos, tan, sec, cosec, and cot) connected with each other? They are connected through trigonometric identities (or in short trig identities). Let's understand the trigonometric identities in detail in the following sections.

Trigonometric identities are very useful for right angles triangles, where you can calculate the value of its sides and angles in just minutes. Moreover, these identities are also useful in practical life situations, for example, calculation of the height of a building etc. Let's find more about such identities in this section.

In algebraic form, an identity in x is satisfied by some particular value of x. For example $(x+1)^2=x^2+2x+1$ is an identity in x. It is satisfied for all values of x. The same applies to trigonometric identities also.

The equations can be seen as facts written in a mathematical form, that is true for "right angle triangle–. Any trigonometric identity dealing with any variable of a right angle triangle will be satisfied by any value within an acceptable range of that variable.

Trigonometry is the branch of mathematics that deals with the relationship between ratios of the sides of a right-angled triangle with its angles. The ratios used to study this relationship are called trigonometric ratios, namely, sine, cosine, tangent, cotangent, secant, cosecant. The word trigonometry is a 16th century Latin derivative and the concept was given by the Greek mathematician Hipparchus.

Here in the below content, we shall understand the basics of trigonometry, the various identities-formulas of trigonometry, and the real-life examples or applications of trigonometry.

INTRODUCTION TO TRIGONOMETRY

Trigonometry is one of the most important branches in mathematics. The word trigonometry is formed by clubbing words 'Trigonon' and 'Metron' which means triangle and measure respectively. It is the study of the relation between the sides and angles of a right-angled triangle. It thus helps in finding the measure of unknown dimensions of a right-angled triangle using formulas and identities based on this relationship.

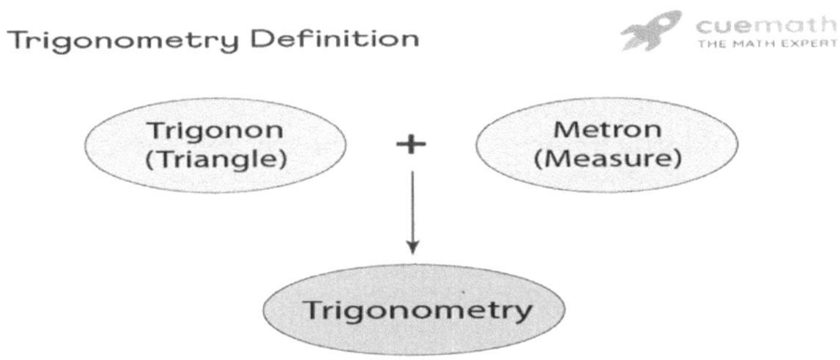

What are Trigonometric Identities?

Trigonometric identities are equations that relate to different trigonometric functions and are true for any value of the variable that is there in the domain. Basically, an identity is an equation that holds true for all the values of the variable(s) present in it.

For example, some of the algebraic identities are:

$(a + b)^2 = a^2 + 2ab + b^2$

$(a - b)^2 = a^2 - 2ab + b^2$

$(a + b)(a-b) = a^2 - b^2$

The algebraic identities relate just the variables whereas the trig identities relate the 6 trigonometric functions sine, cosine, tangent, cosecant, secant, and cotangent. Let's learn about each type of trigonometric identities in detail.

Reciprocal Trigonometric Identities

We already know that the reciprocals of sin, cosine, and tangent are cosecant, secant, and cotangent respectively.

Thus, the reciprocal identities are given as,

- $\sin \theta = 1/\text{cosec}\theta$ (OR) $\text{cosec}\ \theta = 1/\sin\theta$

- $\cos\theta = 1/\sec\theta$ (OR) $\sec\theta = 1/\cos\theta$
- $\tan\theta = 1/\cot\theta$ (OR) $\cot\theta = 1/\tan\theta$

Pythagorean Trigonometric Identities

The Pythagorean trigonometric identities in trigonometry are derived from the Pythagoras theorem. Applying Pythagoras theorem to the right-angled triangle below, we get:

$\text{Opposite}^2 + \text{Adjacent}^2 = \text{Hypotenuse}^2$

Dividing both sides by Hypotenuse^2

$\text{Opposite}^2/\text{Hypotenuse}^2 + \text{Adjacent}^2/\text{Hypotenuse}^2 = \text{Hypotenuse}^2/\text{Hypotenuse}^2$

- $\sin^2\theta + \cos^2\theta = 1$

This is one of the Pythagorean identities. In the same way, we can derive two other Pythagorean trigonometric identities.

- $1 + \tan^2\theta = \sec^2\theta$
- $1 + \cot^2\theta = \text{cosec}^2\theta$

Complementary and Supplementary Trigonometric Identities

The complementary angles are a pair of two angles such that their sum is equal to 90°. The complement of an angle θ is (90 - θ). The trigonometric ratios of complementary angles are:

- $\sin(90° - \theta) = \cos\theta$
- $\cos(90° - \theta) = \sin\theta$
- $\text{cosec}(90° - \theta) = \sec\theta$
- $\sec(90° - \theta) = \text{cosec}\,\theta$
- $\tan(90° - \theta) = \cot\theta$
- $\cot(90° - \theta) = \tan\theta$

The supplementary angles are a pair of two angles such that their sum is equal to 180°. The supplement of an angle θ is (180 - θ). The trigonometric ratios of supplementary angles are:

- $\sin(180° - \theta) = \sin\theta$
- $\cos(180° - \theta) = -\cos\theta$
- $\text{cosec}(180° - \theta) = \text{cosec}\,\theta$
- $\sec(180° - \theta) = -\sec\theta$
- $\tan(180° - \theta) = -\tan\theta$
- $\cot(180° - \theta) = -\cot\theta$

SUM AND DIFFERENCE TRIGONOMETRIC IDENTITES

The sum and difference identities include the formulas of sin(A+B), cos(A-B), cot(A+B), etc.

- $\sin(A+B) = \sin A\cos B + \cos A\sin B$

- sin (A-B) = sin A cos B - cos A sin B
- cos (A+B) = cos A cos B - sin A sin B
- cos (A-B) = cos A cos B + sin A sin B
- tan (A+B) = (tan A + tan B)/(1 - tan A tan B)
- tan (A-B) = (tan A - tan B)/(1 + tan A tan B)

Double and Half Angles Trigonometric Identites

Double angle formulas: The double angle trigonometric identities can be obtained by using the sum and difference formulas.

For example, from the above formulas:

sin (A+B) = sin A cos B + cos A sin B

Substitute A = B = θ on both sides here, we get:

sin (θ + θ) = sinθ cosθ + cosθ sinθ

sin 2θ = 2 sinθ cosθ

In the same way, we can derive the other double-angle identities.

- sin 2θ = 2 sinθ cosθ
- cos 2θ = cos2θ - sin 2θ

  = 2 cos2θ - 1

  = 1 - sin 2 θ
- tan 2θ = (2tanθ)/(1 - tan2θ)

Half Angle Formulas

Using one of the above double angle formulas,

$\cos 2\theta = 1 - 2\sin^2\theta$

$2\sin^2\theta = 1 - \cos 2\theta$

$\sin^2\theta = (1 - \cos 2\theta)/(2)$

sin θ = ±√[(1 - cos 2θ)/2]

Replacing θ by θ/2 on both sides,

sin (θ/2) = ±√[(1 - cos √)/2]

This is the half-angle formula of sin.

In the same way, we can derive the other half-angle formulas.

sin (θ/2) = ±√[(1 - cosθ)/2]

cos (θ/2) = ±√(1 + cosθ)/2

tan (θ/2) = ±√[(1 - cosθ)(1 + cosθ)]

The trigonometric identities that we have learned are derived using the right-angled triangles. There are a few other identities that we use in the case of triangles that are not right-angled.

SINE AND COSINE RULE TRIGONOMETRIC IDENTITIES

The sine rule gives the relation between the angles and the corresponding sides of a triangle. For the non-right-angled triangles, we will have to use the sine rule and the cosine rule. For a triangle with sides 'a', 'b', and 'c' and the respective opposite angles are A, B, and C, sine rule can be given as,

- $a/\sin A = b/\sin B = c/\sin C$
- $\sin A/a = \sin B/b = \sin C/c$
- $a/b = \sin A/\sin B$; $a/c = \sin A/\sin C$; $b/c = \sin B/\sin C$

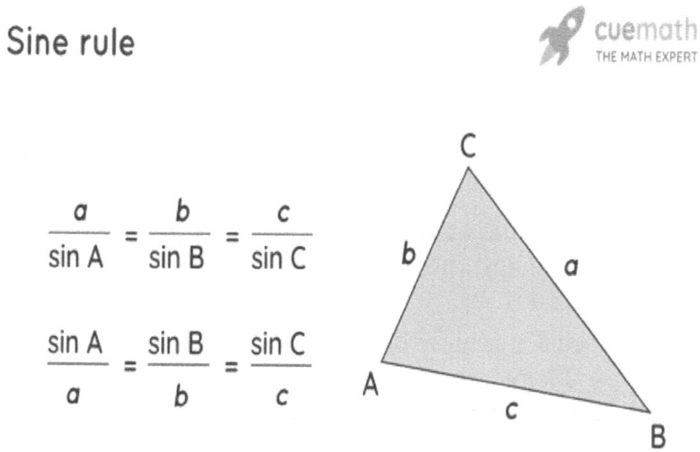

The cosine rule gives the relation between the angles and the sides of a triangle and is usually used when two sides and the included angle of a triangle are given. Cosine rule for a triangle with sides 'a', 'b', and 'c' and the respective opposite angles are A, B, and C, sine rule can be given as,

- $a^2 = b^2 + c^2 - 2bc \cdot \cos A$
- $b^2 = c^2 + a^2 - 2ca \cdot \cos B$
- $c^2 = a^2 + b^2 - 2ab \cdot \cos C$

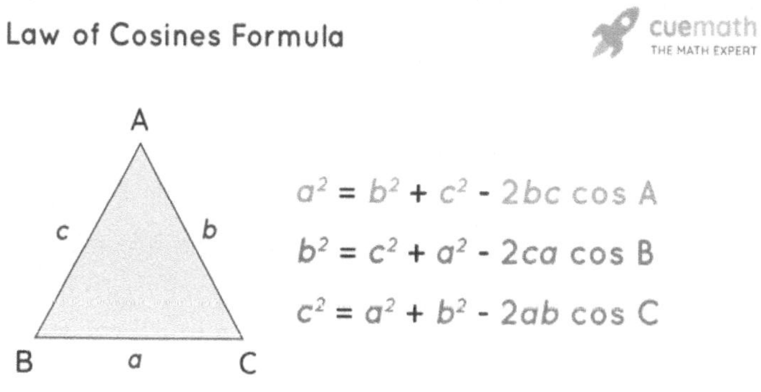

Related Articles on Trigonometric Identities:

- Trigonometric Formula
- Trigonometry
- Inverse Trigonometric Formulas
- Sum and Difference Identities Calculator

Important Notes on Trigonometric Identities

- To write the trigonometric ratios of complementary angles, we consider the following as pairs: (sin, cos), (cosec, sec), and (tan, cot).
- While writing the trigonometric ratios of supplementary angles, the trigonometric ratio won't change. The sign can be decided using the fact that only sin and cosec are positive in the second quadrant where the angle is of the form (180-θ).
- There are 3 formulas for the cos 2x formula. Among them, you can remember just the first one because the other two can be obtained by the Pythagorean identity $\sin^2 x + \cos^2 x = 1$.
- The half-angle formula of tan is obtained by applying the identity tan = sin/cos and then using the half-angle formulas of sin and cos.

Trigonometry Basics

Trigonometry basics deal with the measurement of angles and problems related to angles. There are three basic functions in trigonometry: sine, cosine, and tangent. These three basic ratios or functions can be used to derive other important trigonometric functions: cotangent, secant, and cosecant. All the important concepts covered under trigonometry are based on these functions. Hence, further, we need to learn these functions and their respective formulas at first to understand trigonometry.

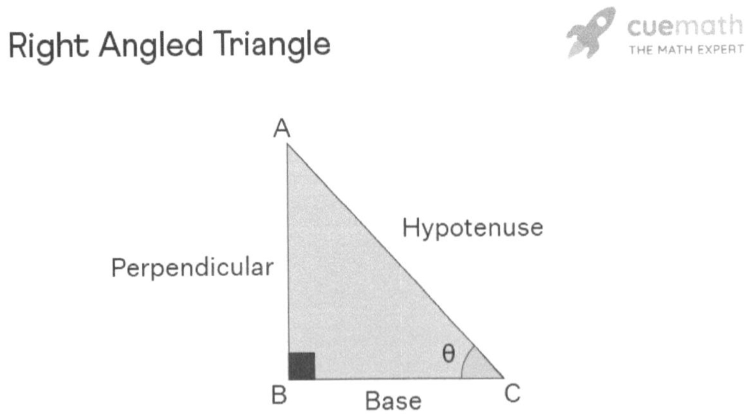

In a right-angled triangle, we have the following three sides.

Perpendicular - It is the side opposite to the angle θ.

Base - This is the adjacent side to the angle θ.

Hypotenuse - This is the side opposite to the right angle.

Trigonometric Ratios

There are basic six ratios in trigonometry that help in establishing a relationship between the ratio of sides of a right triangle with the angle. If θ is the angle in a right-angled triangle, formed between the base and hypotenuse, then

- sin θ = Perpendicular/Hypotenuse
- cos θ = Base/Hypotenuse
- tan θ = Perpendicular/Base

The value of the other three functions: cot, sec, and cosec depend on tan, cos, and sin respectively as given below.

- cot θ = 1/tan θ = Base/Perpendicualr
- sec θ = 1/cos θ = Hypotenuse/Base
- cosec θ = 1/sin θ = Hypotenuse/Perpendicular

Trigonometric Table

The trigonometric table is made up of trigonometric ratios that are interrelated to each other – sine, cosine, tangent, cosecant, secant, cotangent. These ratios, in short, are written as sin, cos, tan, cosec, sec, cot, and are taken for standard angle values. You can refer to the trigonometric table chart to know more about these ratios.

Trigonometric Table

cuemath THE MATH EXPERT

θ	0°	30°	45°	60°	90°	180°	270°	360°
sin θ	0	$\frac{1}{2}$	$\frac{1}{\sqrt{2}}$	$\frac{\sqrt{3}}{2}$	1	0	-1	0
cos θ	1	$\frac{\sqrt{3}}{2}$	$\frac{1}{\sqrt{2}}$	$\frac{1}{2}$	0	-1	0	1
tan θ	0	$\frac{1}{\sqrt{3}}$	1	$\sqrt{3}$	Not Defined	0	Not Defined	0
cosec θ	Not Defined	2	$\sqrt{2}$	$\frac{2}{\sqrt{3}}$	1	Not Defined	-1	Not Defined
sec θ	1	$\frac{2}{\sqrt{3}}$	$\sqrt{2}$	2	Not Defined	-1	Not Defined	1
cot θ	Not Defined	$\sqrt{3}$	1	$\frac{1}{\sqrt{3}}$	0	Not Defined	0	Not Defined

Important Trigonometric Angles

Trigonometric angles are the angles in a right-angled triangle using which different trigonometric functions can be represented. Some standard angles used in trigonometry are 0°, 30°, 45°, 60°, 90°. The trigonometric values for these angles can be observed directly in a trigonometric table. Some other important angles in trigonometry are 180°, 270°, and 360°. Trigonometry angle can be expressed in terms of trigonometric ratios as,

- $\theta = \sin^{-1}$ (Perpendicular/Hypotenuse)
- $\theta = \cos^{-1}$ (Base/Hypotenuse)
- $\theta = \tan^{-1}$ (Perpendicular/Base)

LIST OF TRIGONOMETRIC FORMULAS

There are different formulas in trigonometry depicting the relationships between trigonometric ratios and the angles for different quadrants. The basic trigonometry formulas list is given below:

1. Trigonometry Ratio Formulas

- $\sin\theta$ = Opposite Side/Hypotenuse
- $\cos\theta$ = Adjacent Side/Hypotenuse
- $\tan\theta$ = Opposite Side/Adjacent Side
- $\cot\theta = 1/\tan\theta$ = Adjacent Side/Opposite Side
- $\sec\theta = 1/\cos\theta$ = Hypotenuse/Adjacent Side
- $\operatorname{cosec}\theta = 1/\sin\theta$ = Hypotenuse/Opposite Side

2. Trigonometry Formulas Involving Pythagorean Identities

- $\sin^2\theta + \cos^2\theta = 1$
- $\tan^2\theta + 1 = \sec^2\theta$
- $\cot^2\theta + 1 = \operatorname{cosec}^2\theta$

3. Sine and Cosine Law in Trigonometry

- $a/\sin A = b/\sin B = c/\sin C$
- $c^2 = a^2 + b^2 - 2ab\cos C$
- $a^2 = b^2 + c^2 - 2bc\cos A$
- $b^2 = a^2 + c^2 - 2ac\cos B$

Here a, b, c are the lengths of the sides of the triangle and A, B, and C are the angle of the triangle.

The complete list of trigonometric formulas involving trigonometry ratios and trigonometry identities is listed for easy access. Here's a list of all the trigonometric formulas for you to learn and revise.

Trigonometric Functions Graphs

Different properties of a trigonometric function like domain, range, etc can be studied using the trigonometric function graphs. The graphs of basic trigonometric functions- Sine and Cosine are given below:

Graph of Trignometric Functions

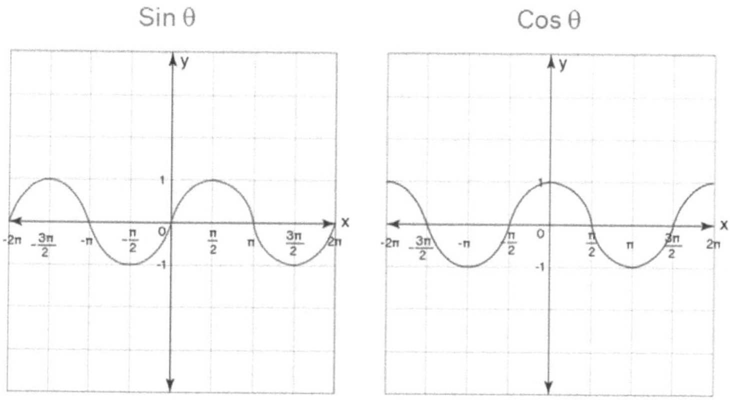

The domain and range of sin and cosine functions can thus be given as,

- sin θ: Domain (-∞, + ∞); Range [-1, +1]
- cos θ: Domain (-∞ +∞); Range [-1, +1]

Click here to learn about the graphs of all trigonometric functions and their domain and range in detail- Trigonometric Functions

Unit Circle and Trigonometric Values

Unit circle can be used to calculate the values of basic trigonometric functions- sine, cosine, and tangent. The following diagram shows how trigonometric ratios sine and cosine can be represented in a unit circle.

Trigonometric Functions Using a Unit Circle

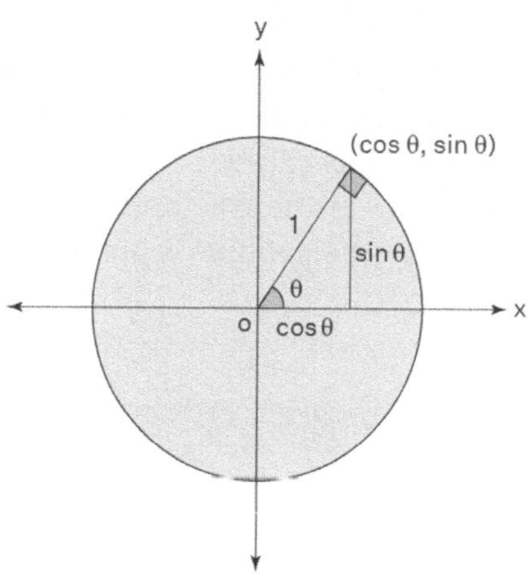

Trigonometry Identities

In Trigonometric Identities, an equation is called an identity when it is true for all values of the variables involved. Similarly, an equation involving trigonometric ratios of an angle is called a trigonometric identity, if it is true for all values of the angles involved. In trigonometric identities, you will get to learn more about the Sum and Difference Identities.

For example, sin θ/cos θ = [Opposite/Hypotenuse] ÷ [Adjacent/Hypotenuse] = Opposite/Adjacent = tan θ

Therefore, tan θ = sin θ/cos θ is a trigonometric identity. The three important trigonometric identities are:

- $\sin^2\theta + \cos^2\theta = 1$
- $\tan^2\theta + 1 = \sec^2\theta$
- $\cot^2\theta + 1 = \text{cosec}^2\theta$

Applications of Trigonometry

Throughout history, trigonometry has been applied in areas such as architecture, celestial mechanics, surveying, etc. Its applications include in:

- Various fields like oceanography, seismology, meteorology, physical sciences, astronomy, acoustics, navigation, electronics, and many more.
- It is also helpful to find the distance of long rivers, measure the height of the mountain, etc.
- Spherical trigonometry has been used for locating solar, lunar, and stellar positions.

REAL-LIFE EXAMPLES OF TRIGONOMETRY

Trigonometry has many real-life examples used broadly. Let's get a better idea of trigonometry with an example. A boy is standing near a tree. He looks up at the tree and wonders "How tall is the tree?" The height of the tree can be found without actually measuring it. What we have here is a right-angled triangle, i.e., a triangle with one of the angles equal to 90 degrees. Trigonometric formulas can be applied to calculate the height of the tree, if the distance between the tree and boy, and the angle formed when the tree is viewed from the ground is given.

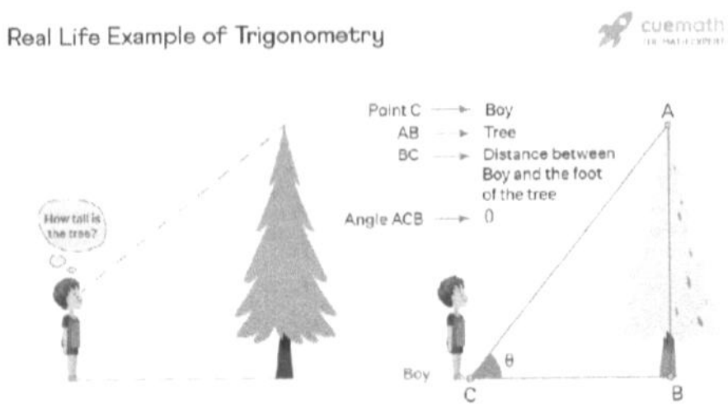

It is determined using the tangent function, such as tan of angle is equal to the ratio of the height of the tree and the distance. Let us say the angle is θ, then

$\tan \theta$ = Height/Distance between object & tree

Distance = Height/$\tan \theta$

Let us assume that distance is 30m and the angle formed is 45 degrees, then

Height = 30/tan 45°

Since, tan 45° = 1

So, Height = 30 m

The height of the tree can be found out by using basic trigonometry formulas.

Sin Cos Tan

In trigonometry, sin, cos, and tan are the basic trigonometric ratios used to study the relationship between the angles and sides of a triangle (especially of a right-angled triangle). Pythagoras worked on the relationship between the sides of a right triangle through the Pythagorean theorem while Hipparcus worked on establishing the relationship between the sides and angles of a right triangle using the concepts of trigonometry. Sin, cos, and tan formulas in trigonometry are used to find the missing sides or angles of a right-angled triangle.

Let's understand the sin, cos, and tan in trigonometry using formulas and examples.

What is Sin Cos Tan in Trigonometry?

Sin, cos, and tan are the three primary trigonometric ratios, namely, sine, cosine, and tangent respectively, where each of which gives the ratio of two sides of a right-angled triangle. We know that the longest side of a right-angled triangle is known as the "hypotenuse" and the other two sides are known as the "legs." That means, in trigonometry, the longest side of a right-angled triangle is still known as the "hypotenuse" but the other two legs are named to be:

- opposite side and
- adjacent side

We decide the "opposite" and "adjacent" sides based upon the angle which we are talking about.

- The "opposite side" or the perpendicular is the side that is just "opposite" to the angle.

- The "adjacent side" or the base is the side(other than the hypotenuse) that "touches" the angle.

  Sin Cos Tan Values

Sin, Cos, and Tan values in trigonometry refer to the values of the respective trigonometric function for the given angle. We can find the sin, cos and tan values for a given right triangle by finding the required ratio of the sides. Let us understand the formulas to find these ratios in detail in the following sections.

SIN COS TAN FORMULAS

Sin, cos, and tan functions in trigonometry are defined in terms of two of the three sides (opposite, adjacent, and hypotenuse) of a right-angled triangle. Here are the formulas of sin, cos, and tan.

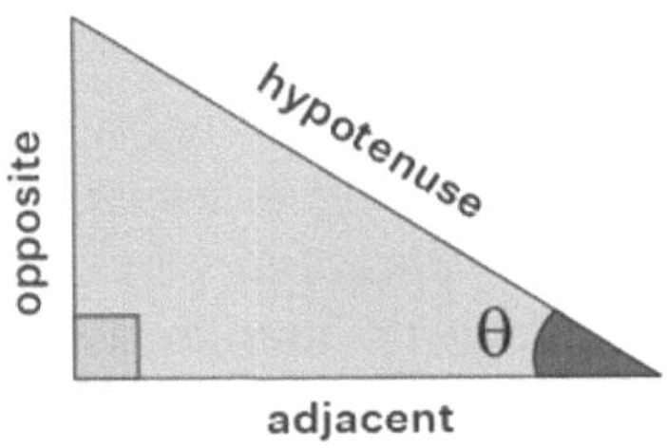

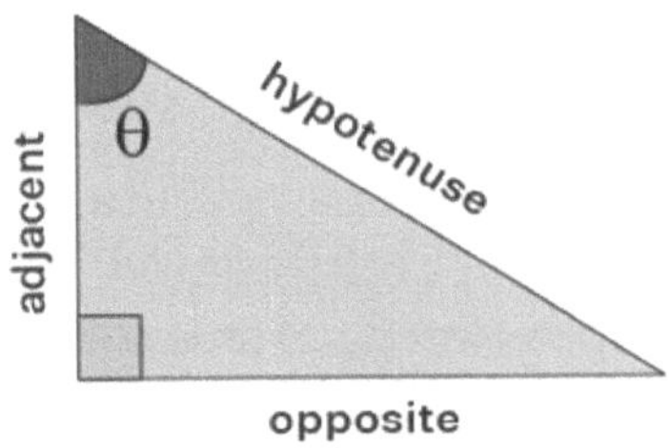

sin θ = Opposite/Hypotenuse

cos θ = Adjacent/Hypotenuse

tan θ = Opposite/Adjacent

Apart from these three trigonometric ratios, we have another three ratios called csc, sec, and cot which are the reciprocals of sin, cos, and tan respectively. Let us understand these sin, cos, and tan formulas using the example given below.

Example: Find the sin, cos, and tan of the triangle for the given angle θ.

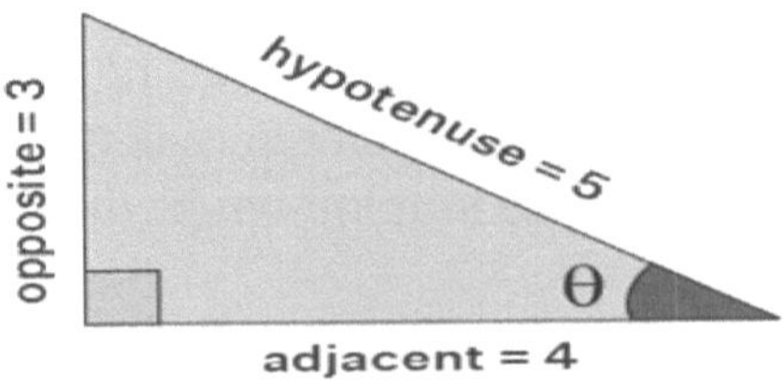

Solution:

In the triangle, the longest side (or) the side opposite to the right angle is the hypotenuse. The side opposite to θ is the opposite side or perpendicular. The side adjacent to θ is the adjacent side or base.

Now we find sin a θ, cosa θ, and tan θ using the above formulas:

sin θ = Opposite/Hypotenuse = 3/5

cos θ = Adjancent/Hypotenuse = 4/5

tan θ = Opposite/Adjacent = 3/4

Trick to remember sin cos tan formulas in trigonometry: Here is a trick to remember the formulas of sin, cos, and tan. We can use the acronym "SOHCAHTOA" as shown below,

$$\text{SOH} \longrightarrow \sin\theta = \frac{\text{Opposite}}{\text{Hypotenuse}}$$

$$\text{CAH} \longrightarrow \cos\theta = \frac{\text{Adjacent}}{\text{Hypotenuse}}$$

$$\text{TOA} \longrightarrow \tan\theta = \frac{\text{Opposite}}{\text{Adjacent}}$$

Sin Cos Tan Table

The trigonometric ratios, sin, cos, and tan do not exactly depend upon the side lengths of the triangle but rather they depend upon the angle because ultimately, we are taking the ratio of the sides. Sin, cos, and tan table is used to find the value of these trigonometric functions for the standard angles. During calculations involving sine, cosine, or tangent ratios, we can directly refer to the trig chart given in the following section to make the deductions easier.

Sin Cos Tan Chart

Sin cos tan chart/table is a chart with the trigonometric values of sine, cosine, and tangent functions for some standard angles 0°, 30°, 45°, 60°, and 90°. We can refer to the trig table given below to directly pick values of sin, cos, and tan values for standard angles.

sin cos tan Chart

θ	0° (or) 0	30° (or) $\frac{\pi}{6}$	45° (or) $\frac{\pi}{4}$	60° (or) $\frac{\pi}{3}$	90° (or) $\frac{\pi}{2}$
sin θ	0	$\frac{1}{2}$	$\frac{\sqrt{2}}{2}$	$\frac{\sqrt{3}}{2}$	1
cos θ	1	$\frac{\sqrt{3}}{2}$	$\frac{\sqrt{2}}{2}$	$\frac{1}{2}$	0
tan θ	0	$\frac{\sqrt{3}}{3}$	1	$\sqrt{3}$	Not Defined

Tips to Remember Sin Cos Tan Table

The tips that you need to memorize from this chart are:

- The angles 0°, 30°, 45°, 60°, and 90° in order.
- The first row (of sin) can be remembered like this: $0/2$, $\sqrt{1}/2$, $\sqrt{2}/2$, $\sqrt{3}/2$.
- That's all you need to remember because:
  The row of cos is as same as the row of sin just in the reverse order.
- Each value in the row of tan is obtained by dividing the corresponding values of sin by cos because tan = sin/cos.
  You can see how is tan = sin/cos here:

$\sin\theta/\cos\theta$ = (Opposite/Hypotenuse) ÷ (Adjacent/Hypotenuse) = (Opposite/Hypotenuse) × (Hypotenuse/Adjacent) = Opposite/Adjacent = $\tan\theta$

Sin Cos Tan on Unit Circle

The values of sin, cos, and tan can be calculated for any given angle using the unit circle. Unit circle in a coordinate plane is a circle of unit radius of 1, frequently centered at the origin (0, 0) in the Cartesian coordinate system in the Euclidean plane, especially in trigonometry. For any point on unit circle, given with the coordinates(x, y), the sin, cos and tan ratios can be given as,

- $\sin\theta = y/1$
- $\cos\theta = x/1$
- $\tan\theta = y/x$

where θ is the angle the line joining the point and origin forms with the positive x-axis.

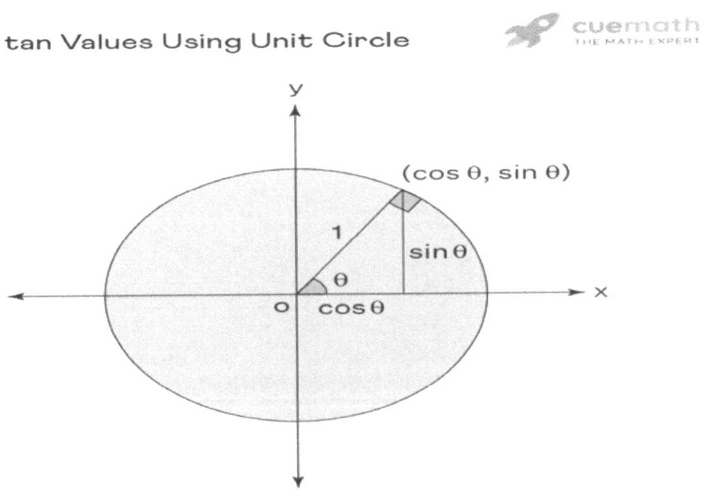

APPLICATION OF SIN COS TAN IN REAL LIFE

Trigonometry ratios sin, cos, tan find application in finding heights and distances in our daily lives. We use sin, cos, and tan to solve many real-life problems. Here is an example to understand the applications of sin, cos and tan.

Example: A ladder leans against a brick wall making an angle of 50° with the horizontal. If the ladder is at a distance of 10 ft from the wall, then up to what height of the wall the ladder reaches?

Solution:

Let us assume that the ladder reaches till x ft of the wall.

Using the given information:

Here, we know the adjacent side (which is 10 ft) and we have to find the opposite side (which is x ft). So we use the relation between the opposite and the adjacent sides which is tan.

$\tan 50^\circ = x/10$

$x = 10 \tan 50^\circ$

$x \approx$" 11.9 ft

Here, tan 50° is calculated using the calculator and the final answer is rounded up to 1 decimal. Therefore, the ladder reaches up to 11.9 ft of the wall.

2

Derivatives of Trig Functions

With this section we're going to start looking at the derivatives of functions other than polynomials or roots of polynomials. We'll start this process off by taking a look at the derivatives of the six trig functions. Two of the derivatives will be derived. The remaining four are left to you and will follow similar proofs for the two given here.

Before we actually get into the derivatives of the trig functions we need to give a couple of limits that will show up in the derivation of two of the derivatives.

FACT

$\lim\theta \to 0\sin\theta/\theta=1\lim\theta \to 0\cos\theta \to 1/\theta=0$

See the Proof of Trig Limits section of the Extras chapter to see the proof of these two limits.

Before proceeding a quick note. Students often ask why we always use radians in a Calculus class. This is the reason why! The proof of the formula involving sine above requires the angles to be in radians. If the angles are in degrees the limit involving sine is not 1 and so the formulas we will derive below would also change. The formulas below would pick up an extra constant that would just get in the way of our work and so we use radians to avoid that. So, remember to always use radians in a Calculus class!

Before we start differentiating trig functions let's work a quick set of limit problems that this fact now allows us to do.

Okay, now that we've gotten this set of limit examples out of the way let's get back to the main point of this section, differentiating trig functions.

We'll start with finding the derivative of the sine function. To do this we will need to use the definition of the derivative. It's been a while since we've had to use this, but sometimes there just isn't anything we can do about it. Here is the definition of the derivative for the sine function.

Trigonometric Function Differentiation

The six trigonometric functions also have differentiation formulas that can be used in application problems of the derivative. The rules are summarized as follows:

1. If f(x) = sin x, then f (x) = cos x
2. If f(x) = cos x, then f (x) = "sin x
3. If f(x) = tan x, then f (x) = sec ^{2} x
4. If f(x) = cot x, then f (x) = "csc ^{2} x.
5. If f(x) = sec x, then f (x) = sec x tan x
6. If f(x) = csc x, then f (x) = "csc x cot x

Note that rules (3) to (6) can be proven using the quotient rule along with the given function expressed in terms of the sine and cosine functions, as illustrated in the following example.

Example 1: Use the definition of the tangent function and the quotient rule to prove if f(x) = tan x, than f (x) = sec ^{2} x.

$$f(x) = \tan x$$

$$= \frac{\sin x}{\cos x}$$

$$f'(x) = \frac{\cos x \cdot \cos x - (-\sin x)\sin x}{\cos^2 x}$$

$$= \frac{\cos^2 x + \sin^2 x}{\cos^2 x}$$

$$= \frac{1}{\cos^2 x}$$

$$= \sec^2 x$$

Example 2: Find y2 if y = x ^{3} cot x.

$$y' = 3x^2 \cot x + x^3(-\csc^2 x)$$

$$= 3x^2 \cot x - x^3 \csc^2 x$$

Example 3: Find if f(x) = 5 sin x + cos x.

$$f'(x) = 5\cos x - \sin x$$

$$f'\left(\frac{\pi}{4}\right) = 5\cos\frac{\pi}{4} - \sin\frac{\pi}{4}$$

$$= \frac{5\sqrt{2}}{2} - \frac{\sqrt{2}}{2}$$

$$= \frac{4\sqrt{2}}{2}$$

$$= 2\sqrt{2}$$

Example 4: Find the slope of the tangent line to the curve y = sin x at the point (π/2,1)

Because the slope of the tangent line to a curve is the derivative, you find that Y = cos x; hence, at (π/2,1), Y = cos π/2 = 0, and the tangent line has a slope 0 at the point (π2,1). Note that the geometric interpretation of this result is that the tangent line is horizontal at this point on the graph of Y = sin x.

One of the most important types of motion in physics is simple harmonic motion, which is associated with such systems as an object with mass oscillating on a spring. Simple harmonic motion can be described by using either sine or cosine functions. In this section we expand our knowledge of derivative formulas to include derivatives of these and other trigonometric functions. We begin with the derivatives of the sine and cosine functions and then use them to obtain formulas for the derivatives of the remaining four trigonometric functions. Being able to calculate the derivatives of the sine and cosine functions will enable us to find the velocity and acceleration of simple harmonic motion.

Differentiation of Trigonometric Functions

The process of finding the derivatives of trigonometric functions is known as the differentiation of trigonometric functions. In other words, the differentiation of trigonometric functions is finding the rate of change of the function with respect to the variable. The six trigonometric functions have differentiation formulas that can be used in various application problems of the derivative.

The six basic trigonometric functions include the following: sine (sin x), cosine (cos x), tangent (tan x), cotangent (cot x), secant (sec x) and cosecant (cosec x). In this article, we will find the derivatives of the trigonometric functions and their proofs. Differentiation of trigonometric functions have applications in different fields such as electronics, computer programming and modeling different cyclic functions.

What is the Differentiation of Trigonometric Functions?

In trigonometry, differentiation of trigonometric functions is a mathematical process of determining the rate of change of the trigonometric functions with respect to the variable angle. The differentiation of trigonometric functions can be done using the derivatives of sin x and cos x by applying the quotient rule. The differentiation formulas of the six trigonometric functions are listed below:

- Derivation of sin x: (sin x)′ = cos x
- Derivative of cos x: (cos x)′ = -sin x
- Derivative of tan x: (tan x)′ = $\sec^2$ x
- Derivative of cot x: (cot x)′ = -cosec^2 x
- Derivative of sec x: (sec x)′ = sec x.tan x
- Derivative of cosec x: (cosec x)′ = -cosec x.cot x

We use d/dx to write the derivatives. Here are the tri derivatives using this notation.

Differentiation of Trigonometric Functions

cuemath THE MATH EXPERT

$$\frac{d}{dx}(\sin x) = \cos x$$

$$\frac{d}{dx}(\cos x) = -\sin x$$

$$\frac{d}{dx}(\tan x) = \sec^2 x$$

$$\frac{d}{dx}(\cot x) = -\text{cosec}^2 x$$

$$\frac{d}{dx}(\sec x) = \sec x \tan x$$

$$\frac{d}{dx}(\text{cosec } x) = -\text{cosec } x \cot x$$

Proofs of Trig Derivatives

Now, that we have the differentiation of trigonometric functions (sin x, cos x, tan x, cot x, sec x, cosec x), we will prove and derive the trig derivatives using various methods such as the quotient rule, the first principle of differentiation, and chain rule along with some limit formulas.

DERIVATIVE OF SIN X

We will derive the derivative of sin x using the first principle of differentiation, that is, using the definition of limits. To derive the differentiation of the trigonometric function sin x, we will use the following limit and trigonometric formulas:

- sin (A+B) = sin A cos B + sin B cos A
- limx→0cosx−1x=0limx→0cosa x−1x=0
- limx→0sinxx=1limx→0sina xx=1

Now, we will calculate the differentiation of trigonometric function sin x:

d (sinx) dx =l imh→ 0 sin (x+h)−sinx(x+h)−x=limh→0 sin xcosh+cosx sinhsinxh=limh→0cosh1hsinx+sinhhcosx=(0)sinx+(1)cosx=cosxd(sinx)dx=limh→sin (x+h)−sinx(x+h)−x=limh→0sina xcosa h+cosa xsin h″sina xh=limh→0cos h−1hsin x+sin hhcos x=(0)sin x+(1)cos x=cosa x

Therefore, d(sin x)/dx = cos x

Derivative of cos x

We will derive the derivative of cos x using the first principle of differentiation, that is, using the definition of limits. To derive the differentiation of the trigonometric function cos x, we will use the following limit and trigonometric formulas:

- cos (A + B) = cos A cos B - sin A sin B
- limx→0cosx−1x=0limx→0cosa x−1x=0
- limx→0sinxx=1limx→0sina xx=1

Thus, we have

d(cosx) dx =l imh →0 cos(x+h)− cosx(x+h)−x= limh→ 0cosxcosh−sin x sinh−cosxh = limh→ 0cosh−1hcosx−sinhhsinx= (0)cosx−(1)sinx=−sinxd(cosx)dx=limh→0cos (x+h)−cos x(x+h)−x=limh→0cos xcos h−sin xsin h−cos xh=limh→0cos h−1hcosx−sina hhsin x=(0)cos x−(1)sin x=−sin x

Therefore, d(cos x)/dx = -sin x

Derivative of tan x

We will determine the derivative of tan x using the quotient rule. We will use the following formulas and identities to calculate the derivative:

- (sin x)′ = cos x
- (cos x)′ = -sin x
- tan x = sin x/ cos x
- $\cos^2 x + \sin^2 x = 1$
- sec x = 1/cos x

(tan x)′ = (sin x/cos x)′

$= [(\sin x)' \cos x - (\cos x)' \sin x]/\cos^2 x$

$= [\cos x. \cos x - (-\sin x). \sin x]/\cos^2 x$

$= (\cos^2 x + \sin^2 x)/\cos^2 x$

$= 1/\cos^2 x$

$= \sec^2 x$

Therefore, $d(\tan x)/dx = \sec^2 x$

Derivative of cot x

We will determine the derivative of cot x using the quotient rule. We will use the following formulas and identities to calculate the derivative:

- $(\sin x)' = \cos x$
- $(\cos x)' = -\sin x$
- $\cot x = \cos x/ \sin x$
- $\cos^2 x + \sin^2 x = 1$
- $\text{cosec } x = 1/\sin x$

$(\cot x)' = (\cos x/\sin x)'$

$= [(\cos x)' \sin x - (\sin x)' \cos x]/\sin^2 x$

$= [-\sin x. \sin x - \cos x. \cos x]/\sin^2 x$

$= (-\sin^2 x - \cos^2 x)/\sin^2 x$

$= -1/\sin^2 x$

$= -\text{cosec}^2 x$

Therefore, $d(\cot x)/dx = -\text{cosec}^2 x$

Derivative of sec x

We will determine the derivative of sec x using the chain rule. We will use the following formulas and identities to calculate the derivative:

- $\sec x = 1/\cos x$
- $\tan x = \sin x/\cos x$
- $(\cos x)' = -\sin x$

$(\sec x)' = (1/\cos x)' = (-1/\cos^2 x).(\cos x)'$

$= (-1/\cos^2 x).(-\sin x)$

$= \sin x/\cos^2 x$

$= (\sin x/\cos x).(1/\cos x)$

$= \tan x \sec x$

Therefore, $d(\sec x)/dx = \tan x \sec x$

Derivative of cosec x

We will determine the derivative of cosec x using the chain rule. We will use the following formulas and identities to calculate the derivative:

- cosec x = 1/sin x
- cot x = cos x/sin x
- (sin x)′ = cos x

  (cosec x)′ = (1/sin x)′ = $(-1/\sin^2 x).(\sin x)'$

  = $(-1/\sin^2 x).(\cos x)$

  = $-\cos x/\sin^2 x$

  = -(cos x/sin x).(1/sin x)

  = -cot x cosec x

  Therefore, d(cosec x)/dx = -cot x cosec x

Applications of Differentiation of Trigonometric Functions

The differentiation of trigonometric functions has various applications in the field of mathematics and real life. Some of them are listed below:

- It is used to determine the slope of the tangent line to a trigonometric curve y = f(x).
- It is used to determine the slope of the normal line to a trigonometric curve y = f(x).
- It helps to determine the equation of the tangent line or normal line of a curve.
- Differentiation of trigonometric functions have applications in different fields such as electronics, computer programming and modeling different cyclic functions.
- We use the derivatives of trigonometric functions to determine the maximum and minimum values of particular functions.

  Differentiation of Inverse Trigonometric Functions

The differentiation of inverse trigonometric functions is done by setting the function equal to y and applying implicit differentiation. Let us list the derivatives of the inverse trigonometric functions along with their domains (arcsin x, arccos x, arctan x, arccot x, arcsec x, arccosec x):

- (arcsin x)′ = $1/\sqrt{(1 - x^2)}$, $-1 < x < 1$
- (arccos x)′ = $-1/\sqrt{(1 - x^2)}$, $-1 < x < 1$
- (arctan x)′ = $1/(1 + x^2)$, $-\infty < x < \infty$
- (arccot x)′ = $-1/(1 + x^2)$, $-\infty < x < \infty$
- (arcsec x)′ = $1/|x|\sqrt{(x^2 - 1)}$, $x\ \varepsilon\ (-\infty, -1) \cup (1, \infty)$
- (arccosec x)′ = $-1/|x|\sqrt{(x^2 - 1)}$, $x\ \varepsilon\ (-\infty, -1) \cup (1, \infty)$

ANTI-DIFFERENTIATION OF TRIGONOMETRIC FUNCTIONS

Anti differentiation of trigonometric functions is the reverse process of differentiation of trigonometric functions. This process is also called the integration of trigonometric functions. The list of anti-derivatives of the trigonometric functions is given below as:

- $\int \sin x \, dx = -\cos x + C$
- $\int \cos x \, dx = \sin x + C$
- $\int \tan x \, dx = \ln |\sec x| + C$
- $\int \cot x \, dx = \ln |\sin x| + C$
- $\int \sec x \, dx = \ln |\sec x + \tan x| + C$
- $\int \operatorname{cosec} x \, dx = -\ln |\operatorname{cosec} x + \cot x| + C$

Here C is the constant of integration.

Inverse Trigonometric Functions

Inverse trigonometric functions, as a topic of learning, are closely related to the basic trigonometric functions. The domain and the range of the trigonometric functions are converted to the range and domain of the inverse trigonometric functions. In trigonometry, we learn about the relationships between angles and sides in a right-angled triangle. Similarly, we have inverse trigonometry functions. The basic trigonometric functions are sin, cos, tan, cosec, sec, and cot. The inverse trigonometric functions on the other hand are denoted as $\sin^{-1}x$, $\cos^{-1}x$, $\cot^{-1} x$, $\tan^{-1} x$, $\operatorname{cosec}^{-1} x$, and $\sec^{-1} x$.

Inverse trigonometric functions have all the formulas of the basic trigonometric functions, which include the sum of functions, double and triple of a function. Here we shall try to understand the transformation of the trigonometric formulas to inverse trigonometric formulas.

What are Inverse Trigonometric Functions?

Inverse trigonometric functions are the inverse functions relating to the basic trigonometric functions. The basic trigonometric function of $\sin \theta = x$, can be changed to $\sin^{-1} x = \theta$. Here, x can have values in whole numbers, decimals, fractions, or exponents. For $\theta = 30°$ we have $\theta = \sin^{-1} (1/2)$, where θ lies between 0° to 90°. All the trigonometric formulas can be transformed into inverse trigonometric function formulas.

Inverse trigonometric functions are also known as the anti-trigonometric functions/ arcus functions/ cyclometric functions. Inverse trigonometric functions are the inverse functions of the basic trigonometric functions which are sine, cosine, tangent, cotangent, secant, and cosecant functions. The inverse trigonometric functions are written using arc-prefix like arcsin(x), arccos(x), arctan(x), arccsc(x), arcsec(x), arccot(x). The inverse trigonometric functions are used to find the angle of a triangle from any of the trigonometric functions. It is used in diverse fields like geometry, engineering, physics, etc.

Consider, the function $y = f(x)$, and $x = g(y)$ then the inverse function can be written as $g = f^{-1}$,

This means that if $y = f(x)$, then $x = f^{-1}(y)$.

An example of inverse trigonometric function is $x = \sin^{-1}y$.

Inverse Trigonometric Formulas

The list of inverse trigonometric formulas has been grouped under the following formulas. These formulas are helpful to convert one function to another, to find the principal angle values of the functions, and to perform numerous arithmetic operations across these inverse trigonometric functions. Further all the basic trigonometric function formulas have been transformed to the inverse trigonometric function formulas and are classified here as the following four sets of formulas.

- Arbitrary Values
- Reciprocal and Complementary functions
- Sum and difference of functions
- Double and triple of a function

Inverse Trigonometric Function Formulas for Arbitrary Values

The inverse trigonometric function formula for arbitrary values is applicable for all the six trigonometric functions. For the inverse trigonometric functions of sine, tangent, cosecant, the negative of the values are translated as the negatives of the function. And for functions of cosine, secant, cotangent, the negatives of the domain are translated as the subtraction of the function from the Π value.

- $\sin^{-1}(-x) = -\sin^{-1}x, x \varepsilon [-1,1]$
- $\tan^{-1}(-x) = -\tan^{-1}x, x \varepsilon R$
- $\text{cosec}^{-1}(-x) = -\text{cosec}^{-1}x, x \varepsilon R - (-1,1)$
- $\cos^{-1}(-x) = \pi - \cos^{-1}x, x \varepsilon [-1,1]$
- $\sec^{-1}(-x) = \pi - \sec^{-1}x, x \varepsilon R - (-1,1)$
- $\cot^{-1}(-x) = \pi - \cot^{-1}x, x \varepsilon R$

Inverse Trigonometric Function Formulas for Reciprocal Functions

The inverse trigonometric function for reciprocal values of x converts the given inverse trigonometric function into its reciprocal function. This follows from the trigonometric functions where sin and cosecant are reciprocal to each other, tangent and cotangent are reciprocal to each other, and cos and secant are reciprocal to each other.

The inverse trigonometric formula of inverse sine, inverse cosine, and inverse tangent can also be expressed in the following forms.

- $\sin^{-1}x = \text{cosec}^{-1}1/x, x \varepsilon R - (-1,1)$
- $\cos^{-1}x = \sec^{-1}1/x, x \varepsilon \quad R - (-1,1)$
- $\tan^{-1}x = \cot^{-1}1/x, x > 0$

  $\tan^{-1}x = -\pi + \cot^{-1}x, x < 0$

Inverse Trigonometric Function Formulas for Complementary Functions

The sum of the complementary inverse trigonometric functions results in a right angle. For the same values of x, the sum of complementary inverse trigonometric functions is equal to a right angle. Therefore, complementary functions of sine-cosine, tangent-cotangent, secant-cosecant, sum up to Π/2. The complementary functions, sine-cosine, tangent-cotangent, and secant-cosecant can be interpreted as,

- $\sin^{-1}x + \cos^{-1}x = \pi/2$, $x \varepsilon [-1,1]$
- $\tan^{-1}x + \cot^{-1}x = \pi/2$, $x \varepsilon R$
- $\sec^{-1}x + \text{cosec}^{-1}x = \pi/2$, $x \varepsilon R - [-1,1]$

Sum and Difference of Inverse Trigonometric Function Formulas

The sum and the difference of two inverse trigonometric functions can be combined to form a single inverse function, as per the below set of formulas. The sum and the difference of the inverse trigonometric functions have been derived from the trigonometric function formulas of sin(A + B), cos(A + B), tan(A + B). These inverse trigonometric function formulas can be used to further derive the double and triple function formulas.

- $\sin^{-1}x + \sin^{-1}y = \sin^{-1}(x.\sqrt{}(1 - y^2) + y\sqrt{}(1 - x^2))$
- $\sin^{-1}x - \sin^{-1}y = \sin^{-1}(x.\sqrt{}(1 - y^2) - y\sqrt{}(1 - x^2))$
- $\cos^{-1}x + \cos^{-1}y = \cos^{-1}(xy - \sqrt{}(1 - x^2).\sqrt{}(1 - y^2))$
- $\cos^{-1}x - \cos^{-1}y = \cos^{-1}(xy + \sqrt{}(1 - x^2).\sqrt{}(1 - y^2))$
- $\tan^{-1}x + \tan^{-1}y = \tan^{-1}(x + y)/(1 - xy)$, if $xy < 1$
- $\tan^{-1}x + \tan^{-1}y = \tan^{-1}(x - y)/(1 + xy)$, if $xy > - 1$

Double of Inverse Trigonometric Function Formulas

The double of an inverse trigonometric function can be solved to form a single trigonometric function as per the below set of formulas.

- $2\sin^{-1}x = \sin^{-1}(2x.\sqrt{}(1 - x^2))$
- $2\cos^{-1}x = \cos^{-1}(2x^2 - 1)$
- $2\tan^{-1}x = \tan^{-1}(2x/1 - x^2)$

These formulas are derived from the double angle formulas of trigonometry.

Triple of Inverse Trigonometric Function Formulas

The triple of the inverse trigonometric functions can be solved to form a single inverse trigonometric function as per the below set of formulas.

- $3\sin^{-1}x = \sin^{-1}(3x - 4x^3)$
- $3\cos^{-1}x = \cos^{-1}(4x^3 - 3x)$
- $3\tan^{-1}x = \tan^{-1}(3x - x^3/1 - 3x^2)$

These formulas resemble and are derived from the triple angle formulas of trigonometry.

Let us study the properties of inverse trigonometric functions using their graph, domain, and range in the following sections.

- Arcsine
- Arccosine
- Arctangent
- Arccotangent
- Arcsecant
- Arccosecant

INVERSE TRIGONOMETRIC FUNCTIONS GRAPH

We can plot the graphs of different inverse trigonometric functions with their range of principal values. Here, we have chosen random values for x in the domain of respective inverse trigonometric functions. We will understand the domain and range of these functions in the following sections.

Arcsine Function

Arcsine function or inverse sine function, also denoted as $\sin^{-1} x$, is an inverse of the sine function. The graph of sine inverse function is as given below.

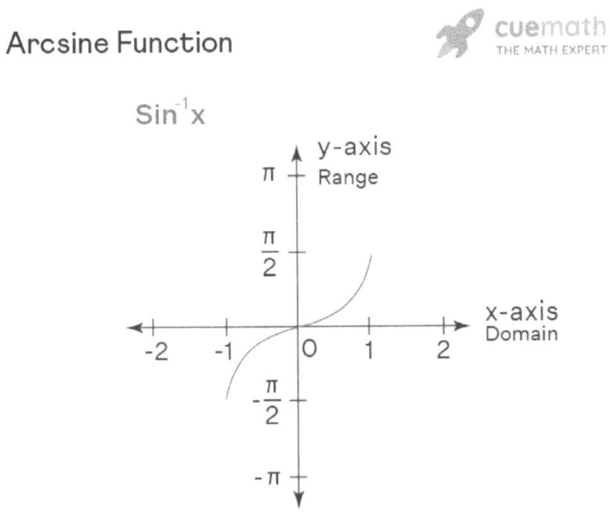

Arccosine Function

Arccosine function or inverse cosine function, also denoted as $\cos^{-1} x$, is an inverse of the cosine function. The graph of cos inverse function is as given below.

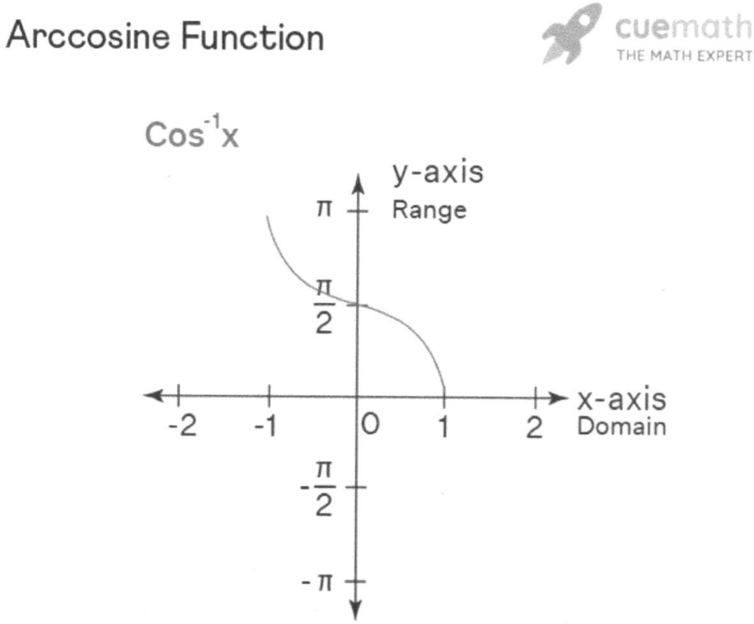

Arctangent Function

Arctangent function or inverse tangent function, also denoted as $\tan^{-1} x$, is an inverse of the tan function. The graph of tan inverse function is as given below.

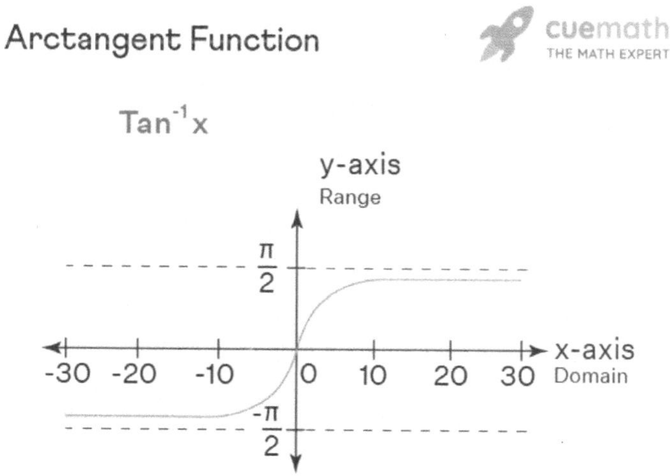

Arccotangent Function

Arccotangent function or inverse cotangent function also denoted as $\cot^{-1} x$, is an inverse of the cotangent function. The graph of cot inverse function is as given below.

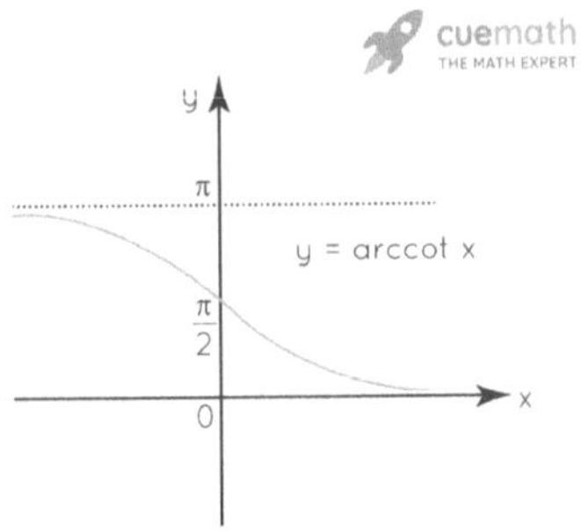

Arcsecant Function

Arcsecant function or inverse secant function, also denoted as $\sec^{-1} x$, is an inverse of the secant function. The graph of sec inverse function is as given below.

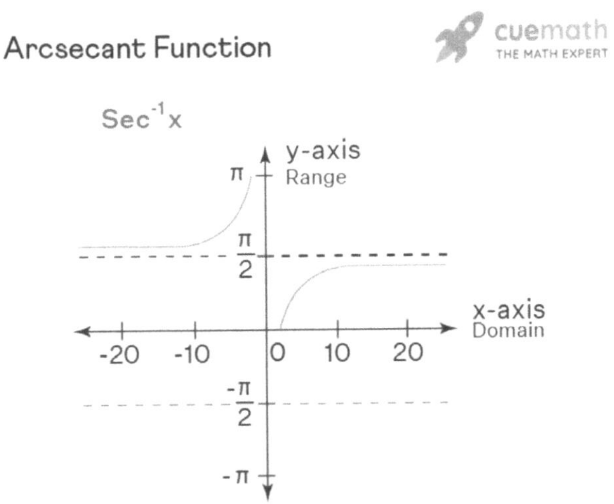

Arccosecant Function

Arccosecant function or inverse cosecant function, also denoted as $\text{cosec}^{-1} x$, is an inverse of the cosecant function. The graph of the cosec inverse function is as given below.

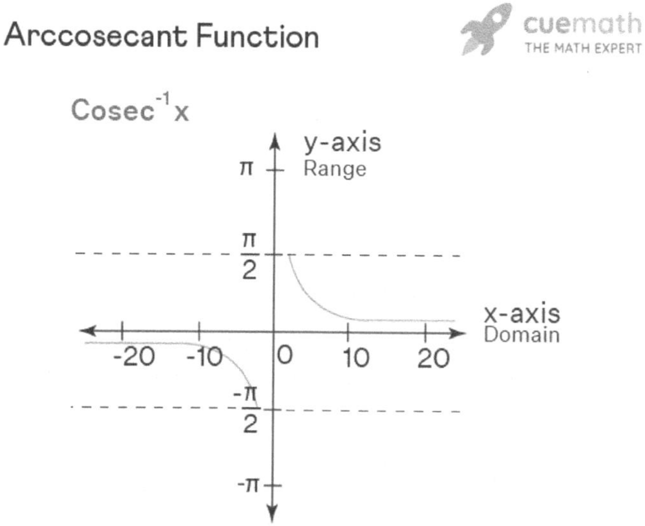

Inverse Trigonometric Functions Domain and Range

The above graphs help us to compare and understand the functions y = $\sin^{-1}x$, y = $\cos^{-1}x$, y = $\tan^{-1}x$, y = $\cot^{-1}x$, y = $\sec^{-1}x$, and y = $\text{cosec}^{-1}x$. The domain(x value) of the function is presented along the x-axis and the range(y value) of the inverse trigonometric function is presented along the y-axis. The below-given table gives the domain and range of principal values of inverse trigonometric functions.

Derivatives of Inverse Trig Functions

We can find the differentiation of different inverse trigonometric functions using differentiation formulas. The following table gives the result of the differentiation of inverse trig functions.

Inverse Trigonometric Functions	dy/dx
$y = \sin^{-1}(x), x \neq -1, +1$	$1/\sqrt{(1-x^2)}$
$y = \cos^{-1}(x), x \neq -1, +1$	$-1/\sqrt{(1-x^2)}$
$y = \tan^{-1}(x), x \neq -i, +i$	$1/(1+x^2)$
$y = \cot^{-1}(x), x \neq -i, +i$	$-1/(1+x^2)$
$y = \sec^{-1}(x), \lvert x\rvert > 1$	$1/[\lvert x\rvert\sqrt{(x^2-1)}]$
$y = \csc^{-1}(x), \lvert x\rvert > 1$	$-1/[\lvert x\rvert\sqrt{(x^2-1)}]$

Integrals of Inverse Trig Functions

Here are the integral formulas of inverse trigonometric functions. To see how to derive each one of them, click here.

Inverse Trigonometric Function	Integral
$\int \sin^{-1}x\, dx$	$x \sin^{-1}x + \sqrt{(1 - x^2)} + C$
$\int \cos^{-1}x\, dx$	$x \cos^{-1}x - \sqrt{(1 - x^2)} + C$
$\int \tan^{-1}x\, dx$	$x \tan^{-1}x - (1/2) \ln \lvert 1 + x^2 \rvert + C$
$\int \csc^{-1}x\, dx$	$x \csc^{-1}x + \ln \lvert x + \sqrt{(x^2 - 1)} \rvert + C$
$\int \sec^{-1}x\, dx$	$x \sec^{-1}x - \ln \lvert x + \sqrt{(x^2 - 1)} \rvert + C$
$\int \cot^{-1}x\, dx$	$x \cot^{-1}x + (1/2) \ln \lvert 1 + x^2 \rvert + C$

SIN COS TAN

In trigonometry, sin, cos, and tan are the basic trigonometric ratios used to study the relationship between the angles and sides of a triangle (especially of a right-angled triangle). Pythagoras worked on the relationship between the sides of a right triangle through the Pythagorean theorem while Hipparcus worked on establishing the relationship between the sides and angles of a right triangle using the concepts of trigonometry. Sin, cos, and tan formulas in trigonometry are used to find the missing sides or angles of a right-angled triangle.

Let's understand the sin, cos, and tan in trigonometry using formulas and examples.

What is Sin Cos Tan in Trigonometry?

Sin, cos, and tan are the three primary trigonometric ratios, namely, sine, cosine, and tangent respectively, where each of which gives the ratio of two sides of a right-angled triangle. We know that the longest side of a right-angled triangle is known as the "hypotenuse" and the other two sides are known as the "legs." That means, in trigonometry, the longest side of a right-angled triangle is still known as the "hypotenuse" but the other two legs are named to be:

- opposite side and
- adjacent side

We decide the "opposite" and "adjacent" sides based upon the angle which we are talking about.

- The "opposite side" or the perpendicular is the side that is just "opposite" to the angle.
- The "adjacent side" or the base is the side(other than the hypotenuse) that "touches" the angle.

Sin Cos Tan Values

Sin, Cos, and Tan values in trigonometry refer to the values of the respective trigonometric function for the given angle. We can find the sin, cos and tan values for a given right triangle by finding the required ratio of the sides. Let us understand the formulas to find these ratios in detail in the following sections.

Sin Cos Tan Formulas

Sin, cos, and tan functions in trigonometry are defined in terms of two of the three sides (opposite, adjacent, and hypotenuse) of a right-angled triangle. Here are the formulas of sin, cos, and tan.

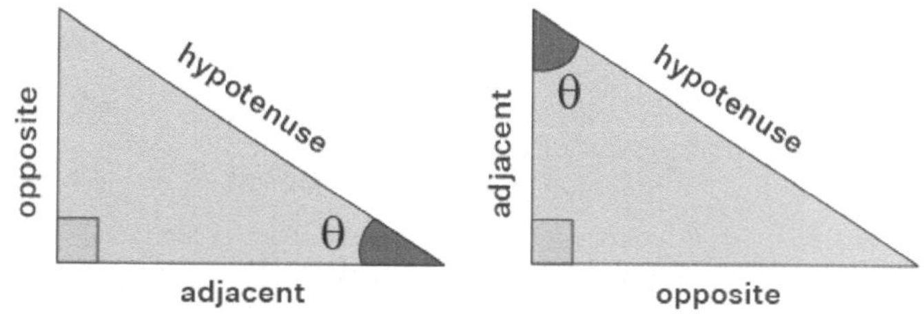

sin θ = Opposite/Hypotenuse

cos θ = Adjacent/Hypotenuse

tan θ = Opposite/Adjacent

Apart from these three trigonometric ratios, we have another three ratios called csc, sec, and cot which are the reciprocals of sin, cos, and tan respectively. Let us understand these sin, cos, and tan formulas using the example given below.

Example: Find the sin, cos, and tan of the triangle for the given angle θ.

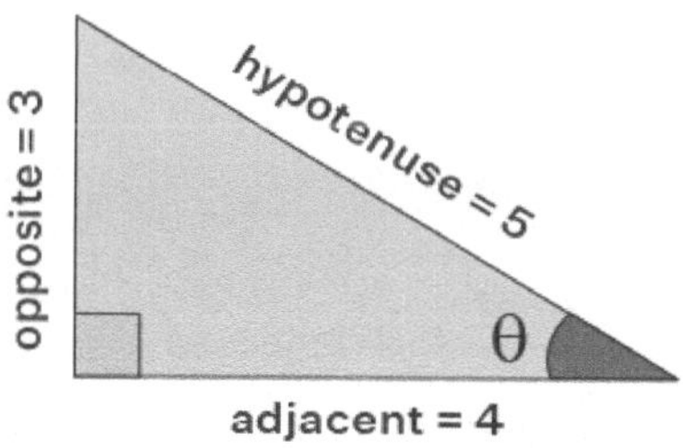

Solution:

In the triangle, the longest side (or) the side opposite to the right angle is the hypotenuse. The side opposite to θ is the opposite side or perpendicular. The side adjacent to θ is the adjacent side or base.

Now we find sin a θ, cosa θ, and tan θ using the above formulas:

sin θ = Opposite/Hypotenuse = 3/5

cos θ = Adjancent/Hypotenuse = 4/5

tan θ = Opposite/Adjacent = 3/4

Trick to remember sin cos tan formulas in trigonometry: Here is a trick to remember the formulas of sin, cos, and tan. We can use the acronym "SOHCAHTOA" as shown below,

$$\text{SOH} \longrightarrow \sin\theta = \frac{\text{Opposite}}{\text{Hypotenuse}}$$

$$\text{CAH} \longrightarrow \cos\theta = \frac{\text{Adjacent}}{\text{Hypotenuse}}$$

$$\text{TOA} \longrightarrow \tan\theta = \frac{\text{Opposite}}{\text{Adjacent}}$$

SIN COS TAN TABLE

The trigonometric ratios, sin, cos, and tan do not exactly depend upon the side lengths of the triangle but rather they depend upon the angle because ultimately, we are taking the ratio of the sides. Sin, cos, and tan table is used to find the value of these trigonometric functions for the standard angles. During calculations involving sine, cosine, or tangent ratios, we can directly refer to the trig chart given in the following section to make the deductions easier.

Sin Cos Tan Chart

Sin cos tan chart/table is a chart with the trigonometric values of sine, cosine, and tangent functions for some standard angles 0°, 30°, 45°, 60°, and 90°. We can refer to the trig table given below to directly pick values of sin, cos, and tan values for standard angles.

sin cos tan Chart

θ	0° (or) 0	30° (or) $\frac{\pi}{6}$	45° (or) $\frac{\pi}{4}$	60° (or) $\frac{\pi}{3}$	90° (or) $\frac{\pi}{2}$
$\sin\theta$	0	$\frac{1}{2}$	$\frac{\sqrt{2}}{2}$	$\frac{\sqrt{3}}{2}$	1
$\cos\theta$	1	$\frac{\sqrt{3}}{2}$	$\frac{\sqrt{2}}{2}$	$\frac{1}{2}$	0
$\tan\theta$	0	$\frac{\sqrt{3}}{3}$	1	$\sqrt{3}$	Not Defined

Tips to Remember Sin Cos Tan Table

The tips that you need to memorize from this chart are:

- The angles 0°, 30°, 45°, 60°, and 90° in order.
- The first row (of sin) can be remembered like this: 0/2, √1/2, √2/2, √3/2.
- That's all you need to remember because:
  The row of cos is as same as the row of sin just in the reverse order.
- Each value in the row of tan is obtained by dividing the corresponding values of sin by cos because tan = sin/cos.
  You can see how is tan = sin/cos here:

sin θ/cos θ = (Opposite/Hypotenuse) ÷ (Adjacent/Hypotenuse) = (Opposite/Hypotenuse) × (Hypotenuse/Adjacent) = Opposite/Adjacent = tan θ

SIN COS TAN ON UNIT CIRCLE

The values of sin, cos, and tan can be calculated for any given angle using the unit circle. Unit circle in a coordinate plane is a circle of unit radius of 1, frequently centered at the origin (0, 0) in the Cartesian coordinate system in the Euclidean plane, especially in trigonometry. For any point on unit circle, given with the coordinates(x, y), the sin, cos and tan ratios can be given as,

- sin θ = y/1
- cos θ = x/1
- tan θ = y/x

where θ is the angle the line joining the point and origin forms with the positive x-axis.

sin cos tan Values Using Unit Circle

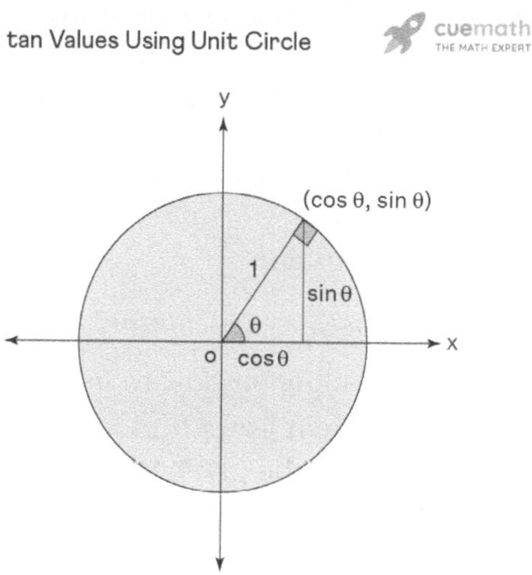

Application of Sin Cos Tan in Real Life

Trigonometry ratios sin, cos, tan find application in finding heights and distances in our daily lives. We use sin, cos, and tan to solve many real-life problems. Here is an example to understand the applications of sin, cos and tan.

Example: A ladder leans against a brick wall making an angle of 50° with the horizontal. If the ladder is at a distance of 10 ft from the wall, then up to what height of the wall the ladder reaches?

Solution:

Let us assume that the ladder reaches till x ft of the wall.

Using the given information:

Here, we know the adjacent side (which is 10 ft) and we have to find the opposite side (which is x ft). So we use the relation between the opposite and the adjacent sides which is tan.

$\tan 50^\circ = x/10$

$x = 10 \tan 50^\circ$

$x \approx$" 11.9 ft

Here, tan 50° is calculated using the calculator and the final answer is rounded up to 1 decimal. Therefore, the ladder reaches up to 11.9 ft of the wall.

Inverse Trig Functions Calculator

Trigonometry is a branch of mathematics that studies relationships between side lengths and angles of triangles.

What is Inverse Trig Functions Calculator?

'Inverse Trig Functions Calculator' is an online tool that helps to calculate the values of the inverse trigonometric function. Online Inverse Trig Functions Calculator helps you to calculate the values of the inverse trigonometric function in a few seconds.

NOTE: Value of 'x' should be between -1 and 1 for sine and cosine.

Inverse Trig Functions Calculator

How to Use Inverse Trig Functions Calculator?

Please follow the steps below on how to use the calculator:

- **Step 1:** Choose a drop-down list to find for $\sin^{-1}$, $\cos^{-1}$, and $\tan^{-1}$.
- **Step 2:** Enter the value x in the given input box.
- **Step 3:** Click on the "Calculate" button to find the values of the inverse trigonometric function

- **Step 4:** Click on the "Reset" button to clear the fields and enter new values.

How to Find Inverse Trig Functions?

The sine function is defined as the ratio of the length of the opposite side to that of the length of the hypotenuse in a right-angled triangle. The sign is a trigonometric function of an angle. It is denoted as $\sin\theta$, where θ is the angle between the two sides.

The inverse sine function is the inverse of the function. If any value x is given, the angle in degrees is calculated for different inverse sine functions. It is denoted by $\sin^{-1}(x)$.

Let function be $y = \sin^{-1}x$. The domain of $\sin^{-1}x$ is $-1 \leq x \leq 1$ and the range of $\sin^{-1}x$ is $-\Pi/2 \leq y \leq \Pi/2$

The cosine function is defined as the ratio of the length of the adjacent side to that of the length of the hypotenuse in a right-angled triangle. Cosine is a trigonometric function of an angle. It is denoted as $\cos\theta$, where θ is the angle between the two sides.

The inverse cosine function is the inverse of the function. If any value x is given, the angle in degrees is calculated for different inverse cosine functions. It is denoted by $\cos^{-1}(x)$

Let function be $y = \cos^{-1}x$. The domain of $\cos^{-1}x$ is $-1 \leq x \leq 1$ and the range of $\cos^{-1}x$ is $0 \leq y \leq \Pi$

The tangent function is defined as the ratio of the length of the opposite side to that of the length of the adjacent side in a right-angled triangle. Tangent is a trigonometric function of an angle. It is denoted as $\tan\theta$, where θ is the angle between the two sides.

The inverse tangent function is the inverse of the function. If any value x is given, the angle in degrees is calculated for different inverse tangent functions. It is denoted by $\tan^{-1}(x)$

Let function be $y = \tan^{-1}x$. The domain of $\tan^{-1}x$ is $-\infty \leq x \leq \infty$ and the range of $\tan^{-1}x$ is $-\Pi/2 \leq y \leq \Pi/2$

The cosecant function is defined as the ratio of the length of the hypotenuse to that of the length of the opposite side in a right-angled triangle. Cosecant is a trigonometric function of an angle. It is denoted as $\operatorname{cosec}\theta$, where θ is the angle between the two sides. It is the inverse of $\sin\theta$.

The inverse cosecant function is the inverse of the function. If any value x is given, the angle in degrees is calculated for different inverse cosecant functions. It is denoted by $\operatorname{cosec}^{-1}(x)$

Let function be $y = \text{cosec}^{-1}x$. The domain of $\text{cosec}^{-1}x$ is $-\infty \le x \le -1$ or $1 \le x \le \infty$ and the range of $\text{cosec}^{-1}x$ is $-\Pi/2 \le y \le \Pi/2,\ y \ne 0$

The secant function is defined as the ratio of the length of the hypotenuse to that of the length of the adjacent side in a right-angled triangle. Secant is a trigonometric function of an angle. It is denoted as $\sec\theta$, where θ is the angle between the two sides. It is the inverse of $\cos\theta$

The inverse secant function is the inverse of the function. If any value x is given, the angle in degrees is calculated for different inverse secant functions. It is denoted by $\sec^{-1}(x)$

Let function be $y = \sec^{-1}x$. The domain of $\sec^{-1}x$ is $-\infty \le x \le -1$ or $1 \le x \le \infty$ and the range of $\sec^{-1}x$ is $0 \le y \le \Pi,\ y \ne \Pi/2$

The cotangent function is defined as the ratio of the length of the adjacent side to that of the length of the opposite side in a right-angled triangle. Cotangent is a trigonometric function of an angle. It is denoted as $\cot\theta$, where θ is the angle between the two sides. It is the inverse of $\tan\theta$.

The inverse cotangent function is the inverse of the function. If any value x is given, the angle in degrees is calculated for different inverse cotangent functions. It is denoted by $\cot^{-1}(x)$

Let function be $y = \cot^{-1}x$. The domain of $\cot^{-1}x$ is $-\infty \le x \le \infty$ and the range of $\cot^{-1}x$ is $0 \le y \le \Pi$

Want to find complex math solutions within seconds?

Use our free online calculator to solve challenging questions. With Cuemath, find solutions in simple and easy steps.

Solved Examples on Inverse Trig Functions Calculator

- **Example1:**
  Find the inverse sine value if the x = 1
  Solution:
  Given x = 1
  $\sin^{-1}(1) = 90°$
- **Example2:**
  Find the inverse tan value if the x = 1
  Solution:
  Given x = 1
  $\tan^{-1}(1) = 45°$

Similarly, you can try the calculator to find the inverse trigonometric functions for the following:

- x = 1/2
- x = -1

3

Trigonometric Ratios

Consider the system of rectangular co-ordinate axes dividing the plane into four quadrants.

A line OP makes angle q with the positive x-axis. The angle q is said to be positive if measured in counter clockwise direction from the positive x-axis and is negative if measured in clockwise direction. The positive values of thetrigonometric ratios in the various quadrants are shown, the signs of the other ratios may be derived.

Note that $\angle xoy = \cap/2$, $\angle xox' = \cap$, $\angle xoy' = 3\cap/2$

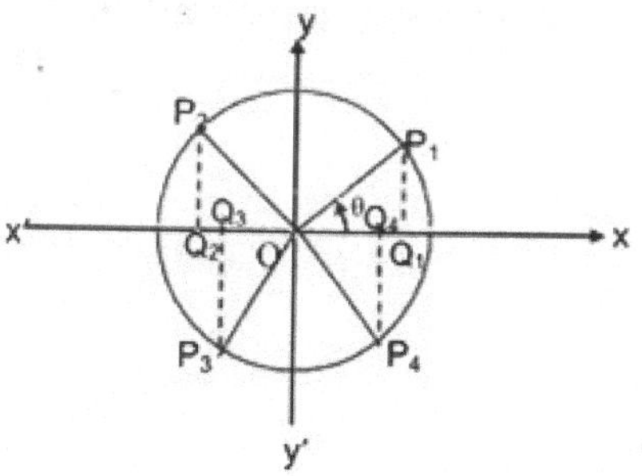

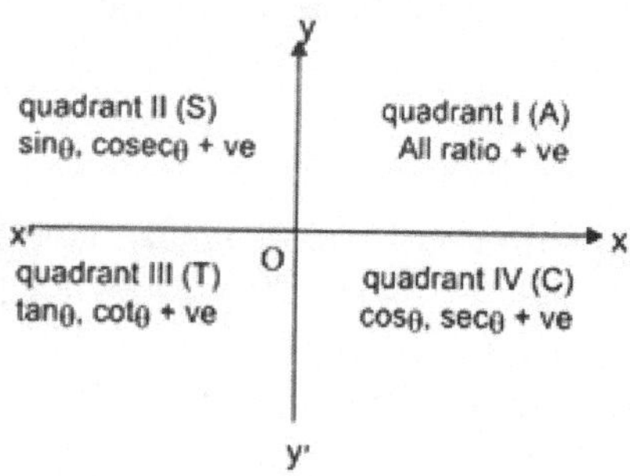

P_iQ_i is positive if above the x-axis, negative if below the x-axis, OP_i is always taken as positive. OQ_i is positive if along the positive x-axis, negative if in the opposite direction.

sin "$Q_iOP_i = P_iQ_i/ OP_i$, cos "Q_iOP_i = OiQi/OQ_i , tan "$Q_iOP_i = P_iQ_i/ OQ_i$ (i = 1, 2, 3).

The depending on signs of OQ_i and P_iQ_i the various trigonometrical ratios will have different signs.

Table – I

a equals	sin a	cos a	tana	cota	seca	coseca
– θ	–sinθ	cosθ	–tanθ	–cotθ	secθ	–cosecθ
90° – θ	cosθ	sinθ	cotθ	tanθ	cosecθ	secθ
90° + θ	cosθ	–sinθ	–cotθ	–tanθ	–cosecθ	secθ
180° – θ	sinθ	–cosθ	–tanθ	–cotθ	–secθ	cosecθ
180° + θ	–sinθ	–cosθ	tanθ	cotθ	–secθ	–cosecθ
360° – θ	–sinθ	cosθ	–tanθ	–cotθ	secθ	–cosecθ
360° + θ	sinθ	cosθ	tanθ	cotθ	secθ	cosecθ

Note:

(i) Angle θ and 90° – θ are complementary angles, θ and 180° – θ are supplementary angles

(ii) $\sin(n\cap + (-1)^n\theta) = \sin\theta$, n ε I

(iii) $\cos(2n\cap + \cap) = \cos\cap$, n ε I

(iv) $\tan(n\cap + \cap) = \tan\cap$, n ε I

i.e. sine of general angle of form $n\cap + (-1)^n\theta$ will have same sign as that of sine of angle θ and so on. The same is true for the respective reciprocal functions also.

Trigonometric Ratios of Compound Angles

An angle made up of the algebraic sum of two or more angles is called a compound angle. Some of the formulae and results regarding compound angles are:

- sin(A + B) = sinA cosB + coA sinB
- sin(A – B) = sinA cosB – cosA sinB
- cos(A + B) = cosA cosB – sinA sinB
- cos(A – B) = cosA cosB + sinA sinB
- tan(A + B) = tanA + tan B/ 1 - tan A tanB
- tan(A – B) = tanA - tan B/ 1 + tan A tan B
- sin(A + B) sin(A – B) = $\sin^2$A – $\sin^2$B = $\cos^2$B – $\cos^2$A.
- cos(A + B) = cos(A – B) = $\cos^2$A – $\sin^2$A – $\sin^2$B = $\cos^2$B – $\sin^2$A.

Illustration:

Prove that tan70° = 2 tan50° + tan20°.

Solution:

tan70° = tan(50° + 20°)

= tan 50° + tan 20°/1 - tan 50°. tan 20°

or, tan70° (1 – tan 50° tan20°) = tan50° + tan20°

or, $\tan 70^\circ = \tan 70^\circ \tan 50^\circ \tan 20^\circ + \tan 50^\circ + \tan 20^\circ$

$= \cot 20^\circ \tan 50^\circ \tan 20^\circ + \tan 50^\circ + \tan 20^\circ$

$= 2 \tan 50^\circ + \tan 20^\circ$.

Illustration:

If $A + B = 45^\circ$, show that $(1 + \tan A)(1 + \tan B) = 2$.

Solution:

$\tan(A + B) = \tan A + \tan B / 1 - \tan A \tan B = 1$

$\tan A + \tan B + \tan A \tan B + 1 = 1 + 1$

$\tan A (1 + \tan B) + (1 + \tan B) = 2$

$(1 + \tan A)(1 + \tan B) = 2$

Trigonometric Ratios of Multiples of an Angle

- $\sin 2A = 2\sin A \cos A = 2\tan A / 1 + \tan^2 A$
- $\cos 2A = \cos 2A - \sin 2A = 1 - 2\sin 2A = 2\cos 2A - 1 = 1 - \tan^2 A / 1 + \tan^2 A$
- $\tan 2A = 2\tan A / 1 - \tan^2 A$
- $\sin 3A = 3\sin A - 4\sin 3A = 4\sin(60^\circ - A) \sin A \sin(60^\circ + A)$
- $\cos 3A = 4\cos 3A - 3\cos A = 4\cos(60^\circ A) \cos A \cos(60^\circ + A)$
- $\tan 3A = 3\tan A - \tan^3 A / 1 - 3\tan^2 A = \tan(60^\circ - A) \tan A \tan(60^\circ + A)$

Illustration:

Find the values of (i) $\sin 18^\circ$ (ii) $\tan 15^\circ$

Solution:

(i) $\sin 18^\circ$

Let $\theta = 18^\circ$ then $2\theta = 36^\circ = 90^\circ - 3\theta$.

Now $\sin 2\theta = 2\sin\theta \cos\theta$ and

$\sin(90^\circ - 3\theta) = \cos^3\theta = 4\cos^3\theta - 3\cos\theta$

Hence we have $2\sin\theta\cos\theta = \cos\theta (4\cos^2\theta - 3) = \cos\theta (1 - 4\sin^2\theta)$.

Hence, $2 \sin\theta = 1 - 4\sin^2\theta$ (as $\cos\theta \# 0$)

$\Rightarrow 4\sin^2\theta + 2\sin\theta - 1 = 0 \Rightarrow \sin\theta = -2 + \sqrt{4} + 16 / 2.4 = -1 + \sqrt{5}/4$

But as $\sin\theta > 0$ we have $\sin\theta = \sqrt{5} - 1/4$ i.e. $\sin 18^\circ = \sqrt{5} + 1/4$

(ii) $\tan 150^\circ$

$\tan 150^\circ = \tan(60^\circ - 45^\circ) = \sqrt{3} - 1/ 1 + \sqrt{3} = (\sqrt{3} - 1)^2/3 - 1 = 4 - 2\sqrt{3}/2 = 2 - \sqrt{3}$.

Sum of sines/cosines in Terms of Products

- $\sin A + \sin B = 2\sin A + B/2 \ \cos A - B/2$
- $\sin A - \sin B = 2\sin A - B/2 \ \cos A + B/2$
- $\cos A + \cos B = 2\cos A - B/2 \ \cos A + B/2$
- $\cos A - \cos B = -2\sin A - B/2 \ \sin A + B/2$ (here notice $(B - A)$)
- $\tan A + \tan B = \sin(A + B) / \cos A . \cos B$

Conversely

- 2sinA cosB = sin(A + B) + sin (A – B)
- 2cosA sinB = sin(A + B) – sin (A – B)
- 2cosA cosB = cos(A + B) + cos (A – B)
- 2sinA sinB = cos(A – B) – cos (A + B)

Illustration:

If a, β and g are in A.P., show that cotβ = sin a - sinγ/ cosγ - cosa .

Solution:

Since a, β and y are in A.P., 2β = a+ y ⇒ cotβ = a+y/2

= (cosa+y/2) / sina+y/2 = (2cosa+y/2 sina+y/2)/ 2sina+y/2sina+y/2 = sin a - siny/ cos y - cosa .

Illustration:

Show that sin 12°.sin48°.sin54° = 1/8.

Solution:

L.H.S. = [cos 36° – cos 60°]sin 54° = 1/2 [cos 36° sin 54° sin 54°]

= 1/4 [2 cos 36° sin 54° – cos 54°] = 1/4 [sin 90° + sin 18° – sin 54°]

= 1/4 [1 – (sin 54°– cos 18°)] = 1/4 [1 – 2sin 18° cos 36°]

= 1/4 [1 - 2sin 18° cos36°] = 1/4 [1 - sin36° cos 36°/ cos18°]

= 1/4 [1 - 2sin 36° cos6°/2cos18°] = 1/4 [1 - sin 72°/2sin72°] = 1/4 [1 - 1/2] = 1/8 = R.H.S.

Alternative Method

Let θ = 12°.

L.H.S. = 1/sin72° sin 12° sin 48° sin 72° sin 54°

= 1/ 4sin 72° 4 sin(60° – 12°) sin 12° sin (60° + 12°) sin 54°.

= 1/4 sin3 (12°) sin 54°/ sin72° = sin 36° sin 54°/ 8 sin 36° cos36° = 1/ 8 = R.H.S.

Trigonometric Ratios of Submultiple of an Angle

- $\left|\sin\frac{A}{2}+\cos\frac{A}{2}\right| = \sqrt{1+\sin A}$

  or $\sin\frac{A}{2}+\cos\frac{A}{2} = \pm\sqrt{1+\sin A}\begin{cases} + & if \quad 2n\pi-\frac{\pi}{4}\le\frac{A}{2}\le 2n\pi+\frac{3\pi}{4} \\ - & otherwise \end{cases}$
- $\left|\sin\frac{A}{2}-\cos\frac{A}{2}\right| = \sqrt{1-\sin A}$

  or $\sin\frac{A}{2}-\cos\frac{A}{2} = \pm\sqrt{1+\sin A}\begin{cases} + & if \quad 2n\pi+\frac{\pi}{4}\le\frac{A}{2}\le 2n\pi+\frac{5\pi}{4} \\ - & otherwise \end{cases}$
- $\tan\frac{A}{2} = \frac{\pm\sqrt{\tan^2 A+1}-1}{\tan A}$

The ambiguities of signs are removed by locating the quadrant in which lies or you can follow the following figure

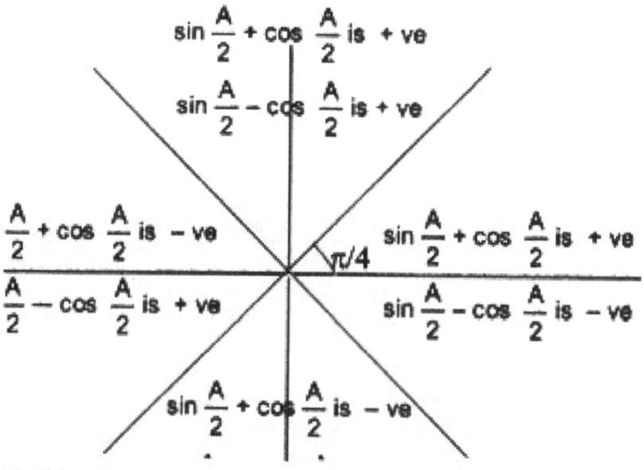

- ,a cosA + b sinA, < $\sqrt{a^2 + b^2}$
  Also cosA + sinA = √2sin (∩/4 + A) = √2cos (A + ‾ ∩/4).

Notes:

- Any formulae that gives the value of sin A/2 in terms of sinA shall also give the value of sine of n∩ +(1)^n A / 2 .
- Any formula that gives the value of cos A/2 in temrs of cosA shall also give the value of cos A 2n∩ + A/2 .
- Any formula that gives the value of tan A/2 in terms of tanA shall also give the value of tan of n∩ +A/2 .

Illustration:

Find the values of sin67 1°/2 and cos67 1°/2 .

Solution:

sin67 1°/2 + cos67 1°/2 = √1 + sin 135°

= √1 + 1/√2 = 1/2 √4 + 2 √2 .... (1)

sin 67 1°/2 - cos 67 1°/2 = √ 1 - sin 135° (because 45° < 67 1°/2 < 135°)

= 1/2 √4 - 2 √2 .... (2)

From (1) and (2), sin67 1°/2 = 1/4 [√4 + 2 √2 + √ 4 - 2 √2] and

cos67 1°/2 = 1/2 [√ 4 + 2 √2 - √4 - 2 √2].

Illustration:

If A, B and C are the angles of a triangle.

Show that tan^2 A/2 + tan^2 B/2 + tan^2 C/2≥ 1

Solution:

We have A/2 + B/2 + C/2 = Π/2 , so that

tan (a/2 + B/2) = tan (Π/2 - C/2) ⇒ tan A/2 + tan B/2 / 1 - tan A/2 tan B/2 = 1/ tan C/2

⇒ tan A/2 tan B/2 + tan B/2 tan C/2 + tan C/2 tan A/2 = 1

$\Rightarrow \tan^2 A/2 \tan^2 B/2 + \tan^2 C/2 - 1 = 1/2 [\Sigma 2 \tan^2 A/2 - \Sigma 2 \tan A/2 \tan B/2]$.

$= 1/2 [(\tan A/2 - \tan B/2)^2 + (\tan B/2 - \tan C/2)^2 + (\tan C/2 - \tan A/2)^2] > 0$

Illustration:

Prove that $(\cos\alpha - \cos\beta)^2 + (\sin\alpha - \sin\beta)^2 = 4\sin^2 (\alpha - \beta/2)$

Solution:

L.H.S. = $\sin \alpha + \sin (\cap - (\cap/3 - \alpha)) + \sin (\cap + \cap/3 + \alpha)$

$= \sin a + \sin (\cap/3 - \alpha) - \sin (\cap/3 + \alpha)$

$= \sin a - [\sin (\cap/3 + \alpha) - \sin (\cap/3 - \alpha)] = \sin\alpha - 2\sin \alpha \cos \cap/3$

Trigonometric Ratios for negative angles

By convention we define an angle as positive if it is measured anticlockwise with the positive X-axis (as shown in figure). Thus any angle measured clockwise from the positive X-axis would be negative. (This is analogous to the case of coordinates in the Cartesian system where for example we take distances to the right of origin as positive distances towards left as negative).

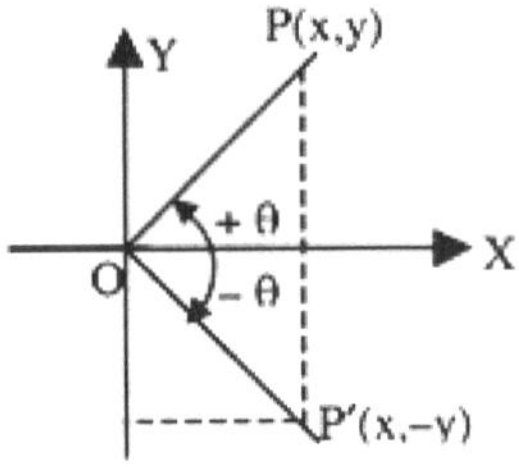

Now, let us establish relation between trigonometric ratio of positive angle and negative angle. From the figure, it is clear that,

$\sin \theta = y/\sqrt{x^2 + y^2}$, and $\sin (-\theta) = (-y)/\sqrt{x^2 + y^2} = -\sin \theta$

$\therefore \sin (-\theta) = -\sin \theta$

Similarly,

$\therefore \cos \theta = x/\sqrt{x^2 + y^2}$, and $\cos (-\theta) = x/\sqrt{x^2 + y^2} = \cos\theta$

$\therefore \cos (-\theta) = \cos \theta$

Properties of Periodicity

1.Constant function is a periodic function without any period. This happens because of the non-existence of the least positive real number happens because of the non-existence of the least positive real number which is due to the continuity of real number system.

2.If f(x) has it's period T then f(ax + b) has its period T/,a,.

3.If f(x) has its period T_1 and g(x) has its period T_2 then (a f(x) + bg(x) has its period ≤ L.C.M. (T_1, T_2). Moreover if f(x) and g(x) are basic trigonometric functions then period of [af (x) + bg(x)] = L.C.M. (T_1, T_2).

Illustration:

Examine whether sin x is a periodic function or not. If so, find its period.

Solution:

Given f(x) = sin x. Let's assume sin x to be periodic. So, it must have some positive value independent of x say T such that

$f(x + T) = f(x)$

$\Rightarrow \sin(x + T) = \sin x$

$\Rightarrow x + T = n\cap + (-1)^n x$ where n = 0, + 1, + 2 ....

Pause : We will learn to solve the trigonometric equations in module III.

The positive values of T independent of x are given by n ∩ where

n = 2, 4, 6.........

further according to definition for periodic number, it should be least. So, here we have T = 2∩.

Thus, it is proved that sin x is periodic function having periodicity 2∩.

Illustration:

Let f(x) = x – [x] where [x] is the greatest integer less than or equal to x. Find out the periodicity of f(x).

Solution:

Assume f(x + T) = f(x)

$\Rightarrow (x + T) - [x + T] = x - [x]$

$\Rightarrow T = [x + T] - [x]$ = an integer.

Hence least positive value of T independent of x is 1.

Thus f(x) is a periodic function of period 1.

Trigonometric Equation and its Solutions

Trigonometric Equation

An equation involving one or more trigonometric ratios of unknown angle is called atrigonometric equation. A trigonometric equation can be written as

Q1 (sinθ, cosθ, tanθ, cotθ, secθ, cosecθ)

= Q2 (sinθ, cosθ, tanθ, cotθ, secθ, cosecθ)

where Q1 and Q2 are rational functions.

Illcos^{2}x – 4 sin x = 1.

This is trigonometric equation as it is not satisfied for all values of x e.g. (2n+1) ∩/4 do not satisfy the given equation.

All possible values of unknown which satisfy the given equation are called solution of the given equation.

For complete solution "all possible values" satisfying the equation must be obtained.

Solving a Trigonometric equation means finding all sets of values of θ, which satisfy the given equation.

Solution of the equation sin θ=0 other than θ=0

Yes, there are many solutions as we shall see below.

Let OX be the initial line

Let "POX = θ and OP = r

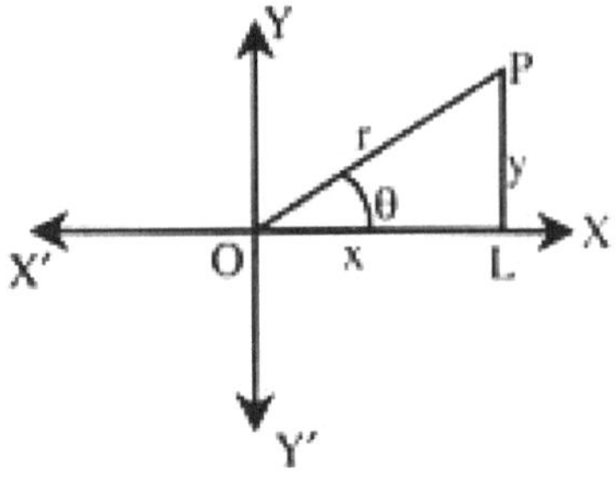

From ΔPOL,

sin θ = PL/OP=y/r.

Now sin θ = 0

⇒ y/r=0; ⇒ y = 0.

This is possible only when OP coincides with OX or OX'.

When OP coincides with OX, θ = 0, ± 2∩, ± 4∩, ± 6∩ .... (1)

And when OP coincides with OX', θ = ±∩, ±3∩, ±5∩ .... (2)

Thus from (1) and (2) it follows that at sinθ = 0

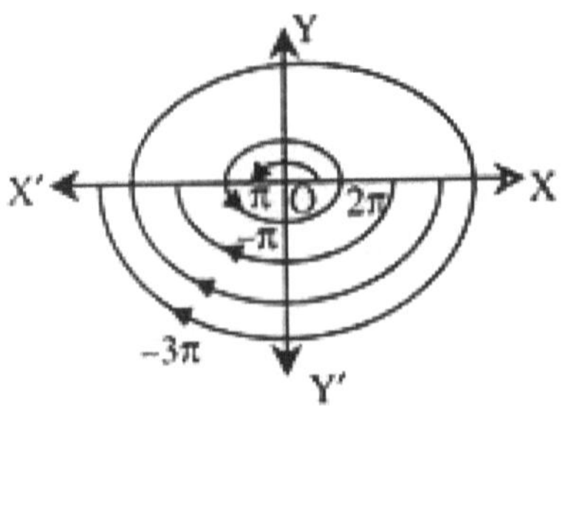

θ = n ∩, where n = 0, ±1, ±2, ...

We call θ = n∩, a general solution of the trigonometric equation sin θ = 0, because for all values of n, this solution satisfies the given equation.

General solution of cos θ=0

cos θ = 0 ⇒ x = ∩/2.

This is possible only when OP coincides with OY or OY′

When OP coincides with OY,

θ = ∩/2,5∩/2,9∩/2 .... (1)

or, -3∩/2-7v∩/2

when OP coincides with OY′

θ = -3∩/2-7∩/2 .... (2)

or, -∩/2-5∩/2

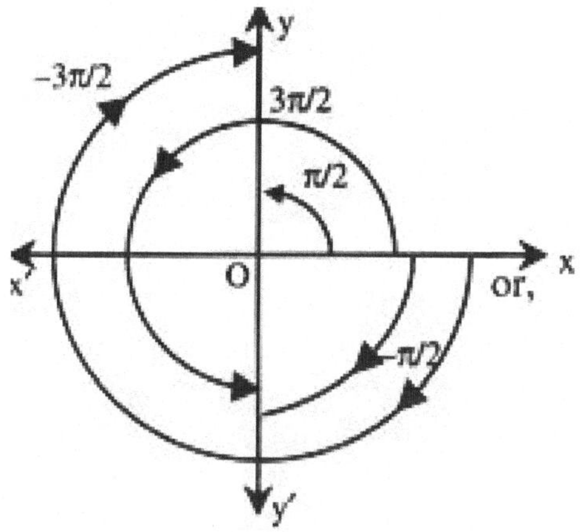

Thus from (1) and (2) if follows that the general solution of cos θ=0 is θ (2n+1) ∩/2, where n = 0, ±1, ±2 ....

General solution of the equation sin θ = k.

We know that when sin θ=k, k has to be such that $-1 \leq k \leq 1$

We can always find some α ε [−∩/2, ∩/2]

As sin (-∩)/2=-1 & sin ∩/2 = 1,

such that sin θ = k, i.e. α = $\sin^{-1}$k

i.e. sin θ = sin α, αε[−∩/2, ∩/2]

⇒ sin θ − sin α = 0

⇒ 2 sin {(θ − α)/2} cos {θ + α)/2} = 0 [we have read it in the last module]

from thc above equation to be satisfied.

either sin {(θ − α)/2) = 0

and consequently (θ − α/2) = integral multiple of ∩

$\therefore\ \theta - \alpha = 2n \cap$

i.e. $\theta = 2n \cap + \alpha$

$\theta = 2n \cap + (-1)^{2n}\ \alpha$ where $n = 0, \pm 1, \pm 2 \ldots$ (1)

or, $\cos \{(\theta + \alpha)/2\} = 0$

i.e. $\{(\theta + \alpha)/2\}$ = Any odd multiple of $\cap/2$

i.e. $\{(\theta + \alpha)/2\} = (2n + 1)\cap/2$

i.e. $\theta = (2n + 1)\cap - \alpha \Rightarrow \theta = (2n + 1) \cap + (-1)^{2n+1}\ \alpha \ldots$ (2)

From (1) and (2), we conclude that

$\theta = n \cap + (-1)n\ \alpha$, where n is integral multiple, is the general solution of the equation $\sin \theta = k$

Illustration:

Find the general solution of equation $\sin \theta = 1/2$

Solution:

We know that $\sin \theta = \sin \cap/6$.

So, general solution of the given equation is $\theta = n \cap + (-1)^n \cap/6$, $n\varepsilon 0, \pm 1, \pm 2$

Important: Following tips and steps will help you systematically solve trigonometric equations.

1. Try to reduce equation in terms of one single trigonometric ratios preferably $\sin \theta$ or $\cos \theta$.

If we have choice to convert a problem in sine or cosine then cosine form is convenient compared to sine form. This is because in general solution of sine, we will have to deal with (–1)n which is inconvenient compared to the dealing of + obtained in cosine form. (See illustraton 5).

2. Factorize the polynomial in terms of these ratio.

3. For LHS to be zero solve for each factor. And write down general solution, for each factor based on the standard results that are derived earlier in this section.

e.g. $\sin\theta - k_1 = 0 \Rightarrow \theta = n \cap + (-1)n \sin^{-1} k_1$

$\cos\theta - k_2 = 0 \Rightarrow \theta = 2n \cap + \cos^{-1} k_2$.

Caution: You must check that k_1, k_2 " [–1, 1]. Don't blindly write as it will be absurd if they do not belong to [–1, 1].

Illustration:

Solve the equation $5 \sin\theta - 2 \cos^2\theta - 1 = 0$

Solution:

Given, $5 \sin\theta - 2 \cos^2\theta - 1 = 0$

or, $5 \sin\theta - 2 (1 - \sin^2\theta) - 1 = 0$

or, $2 \sin^2 \theta + 5 \sin \theta - 3 = 0$

or, $(\sin \theta + 3)(2 \sin \theta - 1) = 0$

$\therefore \sin \theta = 3$ or,

This is not possible as range of sine is $[-1, 1]$

$\sin \theta = 1/2$

or, $\sin \theta = \sin \cap/6$

$\Rightarrow \theta = n \cap + (-1)^n \cap/6$

where $n = 0, \pm 1, \pm 2$ ....

Caution: Never Divide by any expression which is zero. e.g. If the given equation is $(\sin\theta - \cos\theta)(A) = (B)$ where A and B denotes trigonometric equations then you can divide by $(\sin\theta - \cos\theta)$ only when $\theta \neq n \cap + \cap/4$

Illustration:

Solve the equation tan θ + sec θ = "3

Solution:

$\tan \theta + \sec \theta = \sqrt{3}$ ...... (1)

Then $(\sin \theta)/(\cos \theta) + 1/(\cos \theta) = \sqrt{3}$ ...... (2)

or, $\cos(\theta + \cap/6) = \cos \cap/3$

General solution $\theta + \cap/6 = 2n \cap \pm \cap/3, n \in I$

Taking positive sign, $\theta + \cap/6 = 2n \cap + \cap/3$

$\Rightarrow \theta = 2n \cap + \cap/6$

Taking negative sign, $\theta + \cap/6 = 2n \cap - \cap/3$

$\Rightarrow \theta = 2n \cap - \cap/2$

i.e. $\theta = (4n - 1) \cap/2$.

But the solution obtained is correct only if, $\cos \theta \neq 0$ otherwise (2) is not defined.

i.e. $\theta \neq$ odd multiple of $\cap/2$

$\Rightarrow \theta \neq (4n - 1) \cap/2$.

Hence the general solution will be $\theta = 2n \cap + \cap/6$ only where $n = 0, \pm 1, \pm 2$ ......

Caution: Domain of a equation should not change. If it changes, necessary correction must be made in general solution.

Special cases for the Roots

Root loss in an equation

Root loss while solving the equation means that we may miss out some solutions due to some algebraic operations. To avoid this, we should not blindly cancel equal terms on two sides which are in product because this term may also contain additional roots. Cancellation may cause root loss.

Illustration:

$\sin\theta\,[A] = \sin\theta\,[B]$

Solution:

Where A and B are trigonometric functions.

So if we cancel $\sin\theta$ then we loose the solutions, $\sin\theta = 0$

i.e. $\theta = n\cap$, where $n = 0, \pm1, \pm2, \ldots\ldots\ldots$

Extraneous roots

By squaring we may get extra roots. Avoid squaring as far as possible while solving the equation. If squaring is done, do check for extra roots.

Illustration:

$\sec\theta - 1 = (\sqrt{2}-1)\tan\theta$

Solution:

You know that the relation $\sec 2\,\theta = \tan 2\,\theta + 1$. You can very well exploit it by squaring both sides of the given equation is an appropriate manner.

$\Rightarrow \sec\theta = (\sqrt{2}-1)\tan\theta + 1$

or $\sec^2\theta = (\sqrt{2}-1)^2\tan^2\theta + 1 + 2(\sqrt{2}-1)\tan\theta$

or $1 + \tan^2\theta = (2+1-2\sqrt{2})\tan^2\theta + 1 + 2(\sqrt{2}-1)\tan\theta$

or $2(1-\sqrt{2})\tan^2\theta + 2(\sqrt{2}-1)\tan\theta = 0$

or $2(1-\sqrt{2})\tan\theta(\tan\theta - 1) = 0$

either $\tan\theta = 0 \Rightarrow \theta = n\cap$ where $n = 0, \pm1, \pm2\ \ldots\ldots\ldots$

or $\tan\theta - 1 = 0 \Rightarrow \theta = n\cap + \cap/4$. Where $n = 0, \pm1, \pm2, \ldots\ldots$

Note: It is important to note here that all above are not the solutions of the original equation but of its squared version. Squaring many times brings some extra solutions. So we MUST check which of the solutions satisfy the original equations.

Check

(i) for $\theta = n\cap$

$\sec(n\cap) - 1 = (\sqrt{2}-1)\tan(n\cap)$

$(\pm 1) - 1 = 0$

True only when n is even. Hence for odd n, an extra solution has been obtained which does not satisfy the original equation. So, it should be excluded.

$\Rightarrow$ only $\theta = 2k\cap$ is a solution

(ii) for $\theta = n\cap + \cap/4$

$\sec(n\cap + \cap/4) - 1 = (\sqrt{2}-1)\tan(n\cap + \cap/4)$

$\sqrt{2}$-1 = ($\sqrt{2}$-1).

Yes, it does satisfy the given equation. Hence it's a solution

Hence the general solution is

$\theta = (2m \cap), (n \cap + \cap/4)$, where n, m = 0, ±1, ±2, ……

Illustration:

Solve the systems

sin2 x + sin2 y = 1/2.

x – y = 4∩/3

Solution:

We transform the first equation of the system–

1/2 (1 – cos2 x) + 1/2 (1 – cos2y) = 1/2

cos2 x + cos2 y = 1 &rArr 2 cos (x + y).cos(x – y) = 1

Hence it is clear that the system–

cos (x + y) cos (x – y) = 1/2 …… (1)

x – y = 4∩/3 …… (2)

has the same solution as the original system, i.e. the systems are equivalent.

So, from equation (1) and (2) we have

cos (x + y) cos (4∩/3)=1/2 or cos (x + y) = –1

x + y = 2n ∩ ± ∩

Hence we have two linear equations in x and y

x + y = 2n ∩ ± ∩, n ∈ I

x – y = 4∩/3

& rArr x = n ∩ + 2∩/3 ± ∩/2 …… (a)

Taking +ve sign of (a)

x = (n+7/6)∩

= k ∩, k = (n+7/6) and y = k ∩ – 4∩/3

Taking –ve sign of (a)

x = n∩ + ∩/6

y = n∩ + ∩/6-4∩/3

= n ∩ – 7∩/6.

The general solution of system of equations is given by

(k ∩ ,k ∩ -4∩/3) and (n ∩ +∩/6,n ∩ -7∩/2) where k = n + 7/6, n ∈ I

so, these are the solutions to the original system.

Note: In the previous examples we have written the relationships between the unknowns x and the set of solution of the system has been expressed in terms of only one integral parameter e.g. n, k etc. But in practical applications, we may sometimes require to express the general solution in terms of two integral parameters while solving a system of equations with two variables.

Often, you will find that introducing a new variable helps an equation efficiently. Introduction of New Variables can be employed in those cases where a system contains only two trigonometric functions or can be reduced to such a form. Let us see an illustration.

Formulae for solving the trigonometric equations:

- $\sin\theta = \sin\alpha$ and $\cos\theta = \cos\alpha \Rightarrow \theta = 2n\cap + \alpha$
- $\sin\theta = 0 \Rightarrow \theta = n\cap$
- $\cos\theta = 0 \Rightarrow \theta = (2n + 1)\cap/2$
- $\tan\theta = 0 \Rightarrow \theta = n\cap$
- $\sin\theta = \sin\alpha \Rightarrow \theta = n\cap + (-1)^n\alpha$, where $\alpha \in [-\cap/2, \cap/2]$
- $\cos\theta = \cos\alpha \Rightarrow \theta = 2n\cap \pm \alpha$, where $\alpha \in [0, \cap]$
- $\tan\theta = \tan\alpha \Rightarrow \theta = n\cap + \alpha$, where $\alpha \in (-\cap/2, \cap/2)$
- $\sin^2\theta = \sin^2\alpha \cos^2\theta = \cos^2\alpha$, $\tan^2\theta = \tan^2\alpha \Rightarrow \theta = n\cap \pm \alpha$
- $\sin\theta = 1 \Rightarrow \theta = (4n + 1)\cap/2$
- $\sin\theta = -1 \Rightarrow \theta = (4n - 1)\cap/2$
- $\sin\theta = -1 \Rightarrow \theta = (2n + 1)\cap/2$
- $|\sin\theta| = 1 \Rightarrow \theta = 2n\cap$
- $\cos\theta = 1 \Rightarrow \theta = (2n + 1)\cap$
- $|\cos\theta| = 1 \Rightarrow \theta = n\cap$

Note:

- Everywhere in this chapter n is taken as an integer, if not stated otherwise.
- The general solution should be given unless the solution is required in a specified interval or range.
- α is taken as the principal value of the angle. Numerically least angle is called the principal value.

Remarks:

- While solving a trigonometric equation, squaring the equation at any step should be avoided as far as possible and if unavoidable, check the solution for extraneous values.
- Never cancel terms containing unknown terms on the two sides, which are in product. It may cause loss of genuine solution.
- Sometimes the solution of equation may result into more than one solution sets. These solution sets may be disjoint, overlapping or superimposed. If the solution sets are overlapping, we try to rewrite the

solution in such a way that the common solutions are included in one of the solution sets only. If one of the solution set comes out to be the subset of other than we omit the subset form the final solution set.

Important forms of Trigonometric Equations

1. Equations of the form $a \sin^2 x + b \sin x \cos x + c \cos^2 x + d = 0$

If $a + d \neq 0$, then $\cos x = 0$ does not satisfy the equation. We divide the given equation by cos2 x, then equation becomes, $a \tan^2 x + b \tan x + c + d \sec^2 x = 0$.

Writing $\sec^2 x = 1 + \tan^2 x$, we get a quadratic in tan x.

If $a + d = 0$, then the equation becomes

$-d \sin^2 x + b \sin x \cos x + c \cos^2 x + d = 0 \Rightarrow (c + d) \cos^2 x + b \sin x \cos x = 0$

Which can be solved by factorization.

2. Equations of the form $a \sin x + b \cos x + c = 0$

We divide the equation by "(a^2+b^2), thus the equation becomes

$a/\sqrt{(a^2+b^2)} \sin x + b/\sqrt{(a^2+b^2)} \cos x = -c/\sqrt{(a^2+b^2)}$

$\Rightarrow \cos(x-\alpha) = -c/\sqrt{(a^2+b^2)}$, where $a/\sqrt{(a^2+b^2)} = \sin \alpha$ and $b/\sqrt{(a^2+b^2)} = \cos \alpha$

Clearly, the equation has no real solution if $|c| > \sqrt{(a^2+b^2)}$ and if $|c| < \sqrt{(a^2+b^2)}$, we can solve the equation using elementary results.

Alternatively, we can convert the given equation into a quadratic in tan x/2 by writing $\sin x = (2 \tan (x/2))/(1+\tan^2 (x/2))$ and $\cos x = (1-\tan^2 (x/2))/(1+\tan^2(x/2))$.

3. Equations of the form $a (\sin x + \cos x) + b \sin x \cos x + c = 0$

We put $\sin x + \cos x = t \Rightarrow 1 \pm 2 \sin x \cos x = t^2$.

So, the equation reduces to a quadratic equation.

Note:

Due to squaring, the solution must be checked for extraneous roots.

4. Equations containing high degree terms of sin x and / or cos x

We try to convert equation into linear form by using formulae

$\sin^2 x = (1-\cos 2x)/2$, $\cos^2 x=(1+\cos 2x)/2$, $\sin^3 x=(3 \sin x-\sin 3x)/4$, $\cos^3=(3\cos x+\cos 3x)/4$ etc.

Illustration:

Find the general solutions of the following equations

(i) $\tan p\theta + \cot q\theta = 0$

(ii) $\cot \theta + \operatorname{cosec} \theta = \sqrt{3}$

(iii) $5 \sin^? x - 7 \sin x \cos x + 8 \cos 2x + 4 = 0$

(iv) $\sin^{10} x + \cos^{10} x = \cos^4 2x$

Solution:

(i) $\tan p\theta = -\cot q\theta = \tan\left(\frac{\pi}{2} + q\theta\right) \rightarrow p\theta = n\pi + \frac{\pi}{2} - q\theta,\ n \in I$

$\therefore (p + q)\theta = \frac{(2n+1)\pi}{2}$ or $\theta = \frac{(2n+1)\pi}{2(p+q)}$.

(ii) $\cot \theta + \text{cosec}\,\theta - \sqrt{3} \rightarrow \cos \theta + 1 - \sqrt{3}\sin \theta$

or, $\frac{1}{2}\cos\theta - \frac{\sqrt{3}}{2}\sin\theta = -\frac{1}{2} \Rightarrow \cos\left(\theta + \frac{\pi}{3}\right) = \cos\frac{2\pi}{3}$

$\therefore \theta + \frac{\pi}{3} = 2n\pi \pm \frac{2\pi}{3} \Rightarrow \theta = 2n\pi + \frac{\pi}{3}$ or $\theta = (2n - 1)\pi$

If $\theta = (2n - 1)\pi$, then $\cot\theta$ and $\text{cosec}\,\theta$ are undefined

$\therefore \theta = 2n\pi + \frac{\pi}{3}$.

Inverse circular function (Inverse Trigonometric Functions)

Inverse of trigonometric ratios exists.

We know that y = sin x means y is the value of sine of angle x if we consider domain and co-domain both as set R of a real numbers. Sine ratio as seen from the fig. is many-one into function.

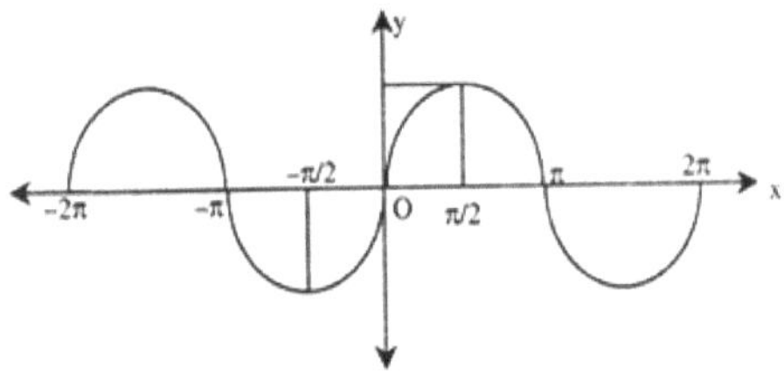

But it is clear that if we restrict the domain to [-∩/2,∩/2] and range to

[–1, 1], then. Y = sin x is one-one onto and hence it is invertible.

So, y = sin x   x ∈ [-∩/2,∩/2], y ∈ [–1, 1]

⇒ x = $\sin^{-1}$ y   y ∈ [–1, 1]. x ∈ [-∩/2,∩/2]

This value of x is called the principle value, i.e. belonging to [-∩/2,∩/2] and [-∩/2,∩/2] range and it is called principal value range.

Note: The smallest numerical angle is called principal value.

In general the inverse circular functions with their domain and range can be tabulated as:

S.No.	Inverse Cir. Fn.	Domain	Range	Graph
1.	$\sin^{-1} x = \theta$ iff $\sin\theta = x, -\pi/2 \le \theta \le \pi/2$	[-1, 1]	[-π/2, π/2]	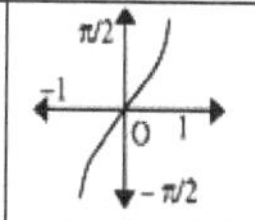
2.	$\cos^{-1} x = \theta$ iff $\cos\theta = x, 0 \le \theta \le \pi$	[-1, 1]	[0, π]	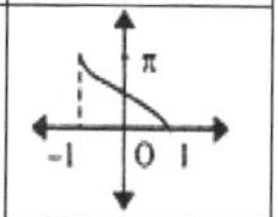
3.	$\tan^{-1} x = \theta$ iff $\tan\theta = x, \frac{\pi}{2} < \theta < \frac{\pi}{2}$	(-∞, ∞)	(-π/2, π/2)	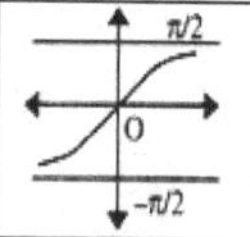
4.	$\cot^{-1} x = \theta$ iff $\cot\theta = x, 0 \le \theta \le \pi$	(-∞, ∞)	(0, π)	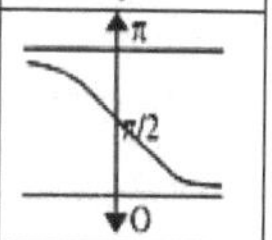
5.	$\sec^{-1} x = \theta$ iff $\sec\theta = x, 0 \le \theta \le \pi$ and $\theta \ne \frac{\pi}{2}$	(-∞, -1] ∪[1, ∞)	[0, π] $\theta \ne \frac{\pi}{2}$	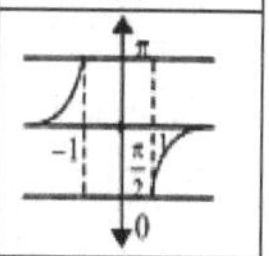
6.	$\text{cosec}^{-1} x = \theta$ iff $\text{cosec}\,\theta = x$ $-\frac{\pi}{2} \le \theta \le \frac{\pi}{2}, \theta \ne 0$	(-∞, -1] ∪[1, ∞)	[-π/2, π/2] $\theta \ne 0$	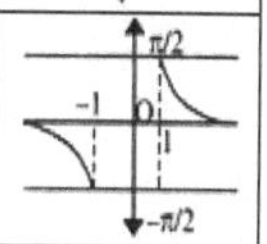

Note: Principal value range of all Inverse Circular function is very important as the function is defined only in this range.

Pause: Odd function are defined as $f(-x) = -f(x)$ and even function as $f(-x) = f(x)$.

The inverse circular functions are defined as below:-

1. $\sin^{-1}(-x) = -\sin^{-1} x$, $\quad -1 < x < 1$ Odd function
2. $\cos^{-1}(-x) = \cap - \cos^{-1} x$, $\quad -1 < x < 1$ Neither odd nor even
3. $\tan^{-1}1(-x) = -\tan^{-1} x$, $\quad x \in R$ Odd function
4. $\cot^{-1}(-x) = \cap - \cot^{-1} x$, $\quad x \in R$ Neither odd nor even
5. $\text{cosec}^{-1}(-x) = -\text{cosec}^{-1} x$, $\quad x < -1$ or $x > 1$ Odd function
6. $\sec^{-1}(-x) = \cap - \sec^{-1} x$, $\quad x < -1$ or $x > 1$ Neither odd nor even

Let us see the proof of any one of the above.

Proof 2: Let $\cos^{-1}(-x) = \theta$, then $\cos\theta = -x$

or, $-\cos\theta = x$ or $\cos(\cap - \theta) = x$

or, $\cap - \theta = \cos^{-1} x$ or $\cos^{-1}(-x) = \cap - \cos^{-1} x$

Similarly we can prove other results.

Caution: Instead of taking cos ($\cap - \theta$) equal to $- \cos \theta$, we could have taken cos ($\cap + \theta$). We opt for cos ($\cap - \theta$) because ($\cap - \theta$) lies in a principal value range i.e. $0 \le \cos^{-1} x \le \cap$.

Illustration:

Find $\cot^{-1} (-1)$

Solution:

$\cot^{-1} (-1) = \cap - \cot^{-1} (1)$

$= \cap - \cap/4$ [5" $\cot^{-1} (-x) = \cap - \cot^{-1} x$]

$= 3\cap/4$.

Properties Of Inverse Circular Function:

1. Self adjusting Properties

(a) $\sin (\sin^{-1} x) = x, x \in [-1, 1]$

Let $\sin^{-1} x = \theta$, θ &isin $(-\cap/2, \cap/2)$ .... (1)

Then $x = \sin \theta$

From (1) putting the value of θ in (2), we get, ... (2)

$\sin (\sin^{-1} x) = \sin (\theta) = x$

(b) $\sin^{-1} \sin x = x, x \in [-\cap/2, \cap/2]$.

Let $\sin x = y, y \in [-1, 1]$ .... (1)

Then $x = \sin^{-1} y$ .... (2)

Putting the value of y in (2) from (1), we get,

$\sin^{-1} \sin x = x$

Illustration:

Evaluate $\sin^{-1} \sin 5$, 5 is radian

Solution:

We know $\sin^{-1} \sin x = x, x \in [-\cap/2, \cap/2]$ or $x \in [-1.57, 1.57]$

We can write, $5 = 2\cap + 5 - 2\cap$

$\sin 5 = \sin (2\cap + (5 - 2\cap))$

$= \sin (5 - 2\cap), 5 - 2\cap \in [-1.57, 1.57]$

$\therefore \sin^{-1} \sin 5 = \sin^{-1} \sin (5 - 2\cap)$

$= 5 - 2\cap$

2. Reciprocal Property

$\sin^{-1} (1/x) = \text{cosec}^{-1} x, x \le -1$, or $x \ge 1$

let $\text{cosec}^{-1} x = \theta \Rightarrow \text{cosec}\, \theta, \theta \in [-\cap/2, \cap/2] - \{0\}$

$\Rightarrow \text{LHS} = \sin^{-1} (1/(\text{cosec}\, \theta)) = \theta$

$\Rightarrow \sin^{-1} (\sin \theta) = \theta$ which is true for above domain of θ

$= \text{cosec}^{-1} x = \text{RHS}$.

Note: $\cot^{-1}(1/x) = \tan^{-1} x$ holds only for $x > 0$. This is because of principal value range of Inverse function cot and tan are different.

When $x < 0$, the principal value range is $\cap/2 < \cot^{-1}(1/x) < \cap$, whereas $-\cap/2 < \tan^{-1}x < 0$. Hence equality will never hold in this case.

$\cot^{-1}(-1/2) = \tan^{-1}(-x)$

$\cap - \cot^{-1}(1/2) = -\tan^{-1} x$

$\cot^{-1}(1/2) = \cap + \tan^{-1} x$

Hence $\cot^{-1}(1/2) = \{\ \tan^{-1} x, \quad x>0$

$\qquad\qquad\qquad\quad \cap+\tan^{-1} x, \quad x<0$

Relation between Inverse trigonometric functions

1. $\sin^{-1} x + \cos^{-1} x = \cap/2 \qquad x \in [-1, 1]$

Let $\sin^{-1} x = \theta$, $\theta \in [-\cap/2, \cap/2]$

$\Rightarrow x = \sin\theta$,

$\Rightarrow x = \cos(\cap/2-\theta), (\cap/2-\theta) \in [0, \cap]$

$\Rightarrow \cos^{-1} x = \cap/2 - \theta$

$\Rightarrow \cos^{-1} x + \sin^{-1} x = \cap/2 \qquad x \in [-1, 1].$

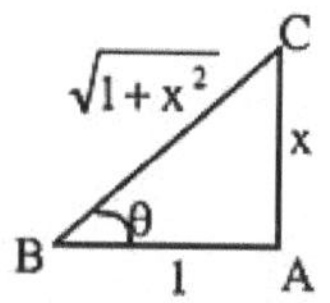

We can express any inverse circular function in terms of the other by drawing a right-angled triangle. Thus, if we have to explain $\tan^{-1} x$ in terms of the other, then take the perpendicular of the triangle as x and base as 1. Now, hypotenuse is $\sqrt{(1+x^2)}$

$\theta = \tan^{-1} x = \cot^{-1} 1/x = \sin^{-1} x/\sqrt{(1+x^2)} = \cos^{-1} 1/\sqrt{(1+x^2)} = \text{cosec}^{-1}\sqrt{(1+x^2)}/x = \sec^{-1}\sqrt{(1+x^2)}$

Illustration:

Prove that $\sin \cot^{-1} \tan \cos^{-1} x = x$.

Solution:

Note: In such problems we proceed with the last term of LHS.

Here, LHS = $\sin \cot^{-1} \tan \cos^{-1} x$

Let $\cos^{-1} x = \theta$, $x = \cos\theta$

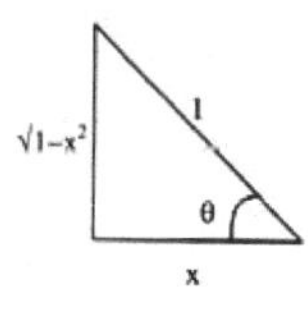

$\therefore \tan\theta = \sqrt{(1+x^2)}/x$

$\tan \cos^{-1} x = \sqrt{(1+x^2)}/x$

(1) Reduces to, $\sin \cot^{-1} (\sqrt{(1+x^2)}/x)$ ...... (2)

Again let $\cot^{-1} \sqrt{(1+x^2)}/x = \varphi$

$\sqrt{(1+x^2)}/x = \cot\varphi$

$\therefore \sin\varphi = x$

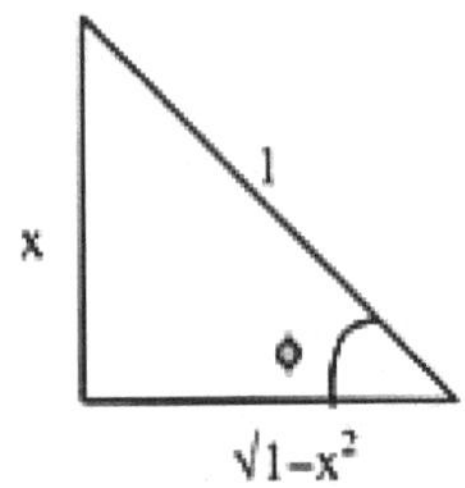

Hence, (2) becomes

$\sin \cot^{-1} \sqrt{(1+x^2)}/x = x = \text{R.H.S.}$

Theorem: $\sin^{-1}x + \sin^{-1}y = \sin^{-1}\left\{\{x\sqrt{(1-y^2)}\} + y\sqrt{(1-x^2)}\right\}.$

$x \geq 0,\ y \geq 0,\ x^2 + y^2 \leq 1$

$\sin^{-1} x + \sin^{-1} y = \pi - \sin^{-1}\left\{\{x\sqrt{(1-y^2)}\} + y\sqrt{(1-x^2)}\right\}$

$x \geq 0,\ y \geq 0,\ x^2 + y^2 > 1$

Proof:

Let $\sin^{-1} x = \alpha$, $\sin^{-1} y = \beta$

$\sin(\alpha + \beta) = \sin\alpha \cos\beta + \cos\alpha \sin\beta$

$= \left\{\{x\sqrt{(1-y^2)}\} + y\sqrt{(1-x^2)}\right\}$

$\Rightarrow \quad \sin^{-1} x + \sin^{-1} y = \alpha + \beta = \sin^{-1}\left\{\{x\sqrt{(1-y^2)}\} + y\sqrt{(1-x^2)}\right\}$

Since $x \geq 0$, $y \geq 0$ so RHS $\in (0, \cap/2)$ but if x and y are such that LHS $> \cap/2$ then above equation is not valid. So above equation is valid only when

$\sin^{-1} x + \sin^{-1} y \leq \cap/2$

$\Rightarrow \sin^{-1} x \leq \cap/2 - \sin^{-1} y$

$\Rightarrow \sin^{-1} x \leq \cos{-1}\, y$

taking sine ( ) both sides

$\sin(\sin^{-1} x) < \sin(\cos{-1}\, y)$

($\therefore$ sine is an increasing function in $[-\cap/2, \cap/2]$)

$\Rightarrow x < \sqrt{(1-y^2)}$

$\Rightarrow x^2 + y^2 < 1$

So above relation is valid only when $x^2 + y^2 \leq 1$

Now, if $x^2 + y^2 > 1$ then

Let us assume LHS = θ and RHS = φ

Then $\sin\theta = \sin\varphi$, which is true only when $\theta = \varphi$ or $\theta = \cap - \varphi$

$\Rightarrow \sin^{-1} x + \sin^{-1} y = \cap - \sin^{-1}(x\sqrt{(1-y^2)} + y\sqrt{(1-x^2)})$.

$\therefore \cos^{-1} x + \cos^{-1} y = \cos^{-1}\{xy - \sqrt{(1-x^2)(1-y^2)}\}$ and $\tan^{-1}\frac{4}{3} = \cos^{-1}\frac{3}{5}$

$\Rightarrow \text{LHS} = \cos^{-1}\left(\frac{12}{13}.\frac{3}{5} - \frac{4}{5}.\frac{5}{13}\right)$

$= \cos^{-1}\left(\frac{36}{65} - \frac{4}{13}\right) = \cos^{-1}\left(\frac{16}{65}\right)$

$= \sin^{-1}\frac{63}{65} = \text{cosec}^{-1} = \frac{65}{63}$ (see adjacent figure)

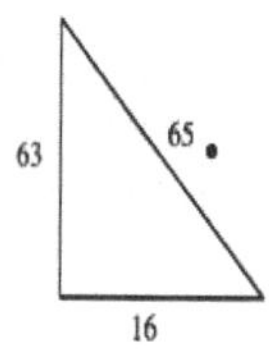

Hence proved.

Theorem: $\tan^{-1} x + \tan^{-1} y = \tan^{-1}\{(x + y)/(1 - xy)\}$ if $xy < 1$

$= \cap - \tan^{-1}(x + y)/(xy - 1)$ if $xy > 1$

$= \cap/2$ if $xy = 1$.

Proof:

Let $\tan^{-1} x = \alpha$ so that $\tan\alpha = x$

and $\tan^{-1} y = \beta$ so that $\tan\beta = y$

Also let $\tan^{-1}((x+y)/(1-xy)) = \gamma$ so that $\tan\gamma = (x+y)/(1-xy)$

We have then to prove that

$\alpha + \beta = \gamma$

Now $\tan(\alpha + \beta) = (\tan\alpha + \tan\beta)/(1 - \tan\alpha\tan\beta) = (x+y)/(1-xy) = \tan\gamma$

So, the relation is proved

$\tan^{-1} x + \tan^{-1} y = \tan^{-1}((x+y)/(1-xy))$ ...... (1)

Case I: When $x \geq 0$, $y \geq 0$ and $xy < 1$, $(x+y)/(1-xy)$ is positive

$\therefore \tan^{-1}[(x+y)/(1-xy)]$ will be positive angle

Case II: When $x \geq 0$, $y \geq 0$ and $xy > 1$, $(x+y)/(1-xy)$ is negative

$\therefore \tan^{-1}((x+y)/(1-xy))$ will be negative angle.

Hence we add ∩ to make it positive

$\therefore \tan^{-1} x + \tan^{-1} y = \cap + \tan^{-1} [(x+y)/(1-xy)]$.

Case III: When xy = 1, (x+y)/(1-xy) = ∞then

$\therefore \tan^{-1} x + \tan^{-1} y = \tan^{-1} \infty = \cap/2$

Illustration:

Show that the equation

tan–1 {((x+1))/((x-1))} + tan–1 {((x-1))/x} = tan–1 (–7) has no solution.

$$\text{LHS} = \tan^{-1}\left\{\frac{\frac{(x+1)}{(x-1)}+\frac{(x-1)}{x}}{1-\frac{(x+1)}{(x-1)}\cdot\frac{(x-1)}{x}}\right\}$$

$$\text{or,}\quad \tan^{-1}\left(\frac{x(x+1)+(x-1)^2}{(x-(x+1))(x-1)}\right)$$

$$\text{or,}\quad \tan^{-1}\left(\frac{x^2+x+x^2-2x+1}{-x+1}\right)$$

$$\Rightarrow \tan^{-1}\left\{\frac{2x^2+1-x}{-x+1}\right\} = \tan^{-1}(-7)$$

$$\Rightarrow \frac{2x^2-x+1}{-x+1} = -7$$

$\Rightarrow (x-2)^2 = 0 \Rightarrow x = 2$

Putting x = 2 in the equation, the equation becomes

tan–1 3 + tan–1 1/2 = tan–1 (–7)

Since x y = 3 . 1/2 > 1

So, tan–1 x + tan–1 y = ∩ + tan–1 ((x+y)/(1-xy))

tan–1 3 + tan–1 1/2 = 3 + tan–1 ((3+1/2)/(1-3/2))

= ∩ + tan–1 (–7) ≠tan–1 (–7) = R.H.S.

Hence given equation has no roots as x = 2 does not satisfy the given equation.

Illustration:

Solve the equation

$\tan^{-1} 2x + \tan^{-1} 3x = n\cap + \cap/4$

Solution:

We know that

$$\tan^{-1} x + \tan^{-1} y = \begin{cases} \tan^{-1}\left(\frac{x+1}{1-xy}\right) & xy < 1 \\ \pi - \tan^{-1}\left\{\frac{x+y}{1-xy}\right\} & xy > 1 \\ \pi/2 & xy = 1 \end{cases}$$

So we have the following cases.

Case I: $2x \,.\, 3x < 1 \Rightarrow x^2 < 1/6 \Rightarrow -1/\sqrt{6} < x < 1/\sqrt{6}$

For this

$\Rightarrow$ L.H.S. = $\tan^{-1} \{(2x+3x)/(1-6x^2)\} = n\cap + \cap/4$

Taking tan of both the sides

$5x/(1-6x^2) = \tan(n\cap + \cap/4) = 1$

$\Rightarrow 6x^2 + 5x - 1 = 0$

$\Rightarrow 6x^2 + 6x - x - 1 = 0$

$\Rightarrow (x + 1)(6x - 1) = 0$

$\Rightarrow x = 1/6$ or $x = -1$

Since $-1 \notin (-1/''6, 1/''6)$

Hence $x = 1/6$ is a solution.

Case II: $2x \,.\, 3x > 1 \Rightarrow x \in (-\infty, -1/\sqrt{6}) \cup (1/\infty 6, \infty)$

For this we have

LHS $= \cap - \tan^{-1}(5x/(6x^2-1)) = n\cap + \cap/4$

$\Rightarrow \tan(\cap - \tan^{-1}(5x/(6x^2-1))) = \tan(n\cap + \cap/4)$

$\Rightarrow -\tan(\tan^{-1}(5x/(6x^2-1))) = 1$

$= 5x = 1 - 6x^2$

$\Rightarrow 6x^2 + 5x - 1 = 0$

$\Rightarrow (x + 1)(x - 1/6) = 0$

$\Rightarrow x = -1, x = 1/6$, since $1/6 \notin (-\infty, -1/\sqrt{6}) \cup (1/\sqrt{6}, \infty)$.

Hence $x = -1$ is another solution.

$\Rightarrow x = -1, 1/6$ are solutions of the given equation.

Theorem: $2\tan^{-1} x = \sin^{-1} 2x/(1+x^2)$, where $|x| < 1$

$= \cos{-1}\,(1-x^2)/(1+x^2)$, where $x > 0$

$= \tan^{-1} 2x/(1-x^2)$, where $|x| < 1$.

Proof: Let $\tan^{-1} x = \theta$, then $\tan\theta = x$

$\sin 2\theta = (2\tan\theta)/(1+\tan^2\theta) = 2x/(1+x^2)$

$\Rightarrow 2\theta = \sin^{-1}(2x/(1+x^2))$

$\Rightarrow 2\tan^{-1} x = \sin^{-1}(2x/(1+x^2))$

Both sides are balanced for $|x| < 1$.

Caution: Restriction on x is $|x| < 1$. To verify it suppose the formulae holds good even for $x > 1$.

Then L.H.S. $= 2\tan^{-1} x$

$= 2\tan^{-1}$ (a no. $1 > 1$) [for $x > 1$]

$> \cap/2$ [$\therefore \tan^{-1}(>1) > \cap/4$], [for principal values]

R.H.S. = sin–1 $2x/(1+x^2)$ which lies between -∩/2 and ∩/2 for all values of x, [for principal values]

Thus, we see that for $x > 1$,

L.H.S. > ∩/2, whereas R.H.S. ≤ ∩/2.

Hence two cannot be equal.

$$\text{We know, } \cos 2\theta = \frac{1-\tan^2\theta}{1+\tan^2\theta} = \frac{1-x^2}{1+x^2}$$

$$2\theta = \cos^{-1}\left(\frac{1-x^2}{1+x^2}\right)$$

$$2\tan^{-1} x = \cos^{-1}\left(\frac{1-x^2}{1+x^2}\right)$$

Both sides are balanced for $x \geq 0$

$$\text{Similarly, } \tan^2\theta = \frac{2\tan\theta}{1-\tan^2\theta} = \frac{2x}{1-x^2}$$

$$2\theta = 2\tan^{-1} x = \tan^{-1}\frac{2x}{1-x^2}$$

Both sides are balanced for $|x| < 1$.

Note: For $|x| > 1$

$2\tan^{-1} x$ = ∩ – sin–1 $2x/(1+x^2)$

= ∩ – cos–1 $(1-x^2)/(1+x^2)$

= ∩ – $\tan^{-1} 2x/(1-x^2)$

SOLVED QUESTIONS

1. Solve the equation: $\tan 5\theta = \tan 3\theta$

Solution:

Now $\tan 5\theta = \tan 3\theta \Rightarrow 5\theta$ = n∩ + 3θ

or, 2θ = n∩

θ = n∩/2, where n = 0, ±1, ±2 ……

putting $n = 0, \theta = 0$, original equation is satisfied

putting $n = 1, \theta = \cap/2$ equation becomes

$\tan 5\cap/2 = \tan 3\cap/2.$

Equation is not defined for odd multiple of $\cap/2$.

Hence we conclude that $\theta = 2n\cap$, where $n = 0, \pm1, \pm2$ ....

Note : 1. If $\tan\theta$ or $\sec\theta$ is involved in the equation, $\theta \neq$odd multiple of $\cap/2$.

2. If $\cot\theta$ or $\operatorname{cosec}\theta$ is involved in the equation, $\theta \neq$multiple of $\cap$ or 0.

2. Solve the system

$\sin x + \cos y = 1$ .... (1)

$\cos2 x - \cos2 y = 1$ .... (2)

Solution:

We can transform the equation (2) by putting $\cos2 x = 1 - 2 \sin2 x$ and $\cos2 y = 2\cos2 y - 1$

$\cos2 x - \cos2 y = 1 - 2\sin^2 x + 1 - 2\cos^2 y = 1$

$\Rightarrow \sin2 x + \cos2 y = 1/2$

and hence our system is

$\sin x + \cos y = 1$ .... (3)

$\sin2 x + \cos^2 y = 1/2$ .... (4)

which is equivalent to the original system.

For simplicity, let us put $\sin x = y$, $\cos y = v$ [note $\rightarrow u \in [-1, 1]$ and $v \in [-1, 1]$] and hence

$u + v = 1$ .... (5)

$u^2 + v^2 = ½$ .... (6)

solving (5) and (6) we get,

$u = \sin x = 1/2$

$v = \cos y = 1/2$

Hence, the general solution of the given system of equations is given by

$x = m\cap + (-1)^m \cap/6, m \in I$

$y = 2n\cap + \cap/3, n \in I$

The solution of various pairs formed from these values of x and y is precisely the set of all solution of the original system.

3. Find the general solutions of the following equations

(i) $\tan p\theta + \cot q\theta = 0$

(ii) $\cot\theta + \operatorname{cosec}\theta = \sqrt{3}$

(iii) $5 \sin^2 x - 7 \sin x \cos x + 8 \cos 2x + 4 = 0$

(iv) $\sin^{10} x + \cos^{10} x = \cos^4 2x$

Solution:

(i) $\tan p\theta = -\cot q\theta = \tan\left(\frac{\pi}{2} + q\theta\right) \rightarrow p\theta = n\pi + \frac{\pi}{2} - q\theta,\ n \in I$

$\therefore (p + q)\theta = \frac{(2n+1)\pi}{2}$ or $\theta = \frac{(2n+1)\pi}{2(p+q)}$.

(ii) $\cot 0 + \operatorname{cos\,ec} 0 - \sqrt{3} \rightarrow \cos 0 + 1 - \sqrt{3} \sin 0$

or, $\frac{1}{2} \cos \theta - \frac{\sqrt{3}}{2} \sin\theta = -\frac{1}{2} \Rightarrow \cos\left(\theta + \frac{\pi}{3}\right) = \cos \frac{2\pi}{3}$

$\therefore \theta + \frac{\pi}{3} = 2n\pi \pm \frac{2\pi}{3} \Rightarrow \theta = 2n\pi + \frac{\pi}{3}$ or $\theta = (2n - 1)\pi$

If $\theta = (2n - 1)\pi$, then $\cot \theta$ and $\operatorname{cosec} \theta$ are undefined

$\therefore \theta = 2n\pi + \frac{\pi}{3}$.

(iii) $5 \sin^2 x - 7 \sin x \cos x + 8 \cos 2x + 4 = 0 \Rightarrow 5 \sin^2 x - 7 \sin x \cos x + 16 \cos^2 x - 4 = 0$

Dividing by $\cos^2 x$, we get

$5 \tan^2 x - 7 \tan x + 16 - 4 (1 + \tan^2 x) = 0$

$\tan x = 3 \Rightarrow x = n\pi + \tan^{-1} 3$

$\tan x = 4 \Rightarrow x = m\pi + \tan^{-1} 4,\ n, m \in I$

(iv) $\sin^{10} x + \cos^{10} x = \cos^4 2x$

$\Rightarrow \left\{\frac{1-\cos 2x}{2}\right\}^5 + \left\{\frac{1+\cos 2x}{2}\right\}^5 = \cos^4 2x$

$\Rightarrow 1 + {}^5C_2 \cos^2 2x + {}^5C_4 \cos^4 2x = 16 \cos^4 2x$

$\Rightarrow 11 \cos^4 2x - 10 \cos^2 2x - 1 = 0$

$\Rightarrow (11 \cos^2 2x + 1)(\cos^2 2x - 1) = 0$

$\Rightarrow \cos^2 2x - 1 = 0 \Rightarrow \sin^2 2x = 0 \Rightarrow 2x = n\pi$ or $x = \frac{n\pi}{2},\ n \in I$.

4. Solve $\sin^2\theta - \cos\theta = ¼$ for θ and write the values of θ in the interval $0 < \theta < 2\cap$.

Solution:

The given equation can be written as

$1 - \cos^2\theta - \cos\theta = 1/4$

$\Rightarrow \cos^2\theta + \cos\theta - 3/4 = 0 \Rightarrow 4\cos^2\theta + 4\cos\theta - 3 = 0$

$\Rightarrow \cos\theta = (-4\pm\sqrt{(16+48)})/8=1/2,-3/2.$

Since $\cos\theta = -3/2$ is not possible,

$\cos\theta = 1/2 = \cos \pi/3 \Rightarrow \theta = 2n\pi \pm \pi/3.$

For the given interval, n = 0 and n = 1 $\Rightarrow \theta = \pi/3, 5\pi/3.$

5. Solve $3\cos^2\theta - 2\sqrt{3}\sin\theta\cos\theta - 3\sin^2\theta = 0.$

Solution:

The given equation can be written as ($\cos\theta \neq 0$).

$3\tan^2\theta + 2\sqrt{3}\tan\theta - 3 = 0$

$\Rightarrow \tan\theta = (-2\sqrt{3}\pm\sqrt{(12+36)})/6=(-2\sqrt{3}\pm4\sqrt{3})/6=1/\sqrt{3},-\sqrt{3}$

Either $\tan\theta = 1/\sqrt{3} = \tan \pi/6 \Rightarrow \theta = n\pi + \pi/6$

or $\tan\theta = -\sqrt{3} = \tan(-\pi/3) \Rightarrow \theta = n\pi - \pi/3.$

6. Solve $\cos\theta + \cos3\theta + \cos5\theta + \cos7\theta = 0.$

Solution:

$\cos\theta + \cos3\theta + \cos5\theta + \cos5\theta + \cos7\theta = 0.$

$\Rightarrow 2\cos4\theta\cos3\theta + 2\cos4\theta\cos\theta = 0$

$\Rightarrow \cos4\theta\,(2\cos2\theta\cos\theta) = 0.$

Either $\cos\theta = 0 \Rightarrow \theta = (2n + 1)\pi/2$

or $\cos2\theta = 0 \Rightarrow \theta = (2n + 1)\pi/4$ or $\cos4\theta = 0$

$\Rightarrow \theta = (2n + 1)\pi/8.$

7. Find the value of $\cos(2\cos^{-1} x + \sin^{-1} x)$ at $x = 1/5$, Where $0 \le \cos^{-1} x \le \pi$, $-\pi/2 \le \sin^{-1} x \le \pi/2$

Solution:

$\cos(2\cos^{-1} x + \sin^{-1} x) = \cos(\cos^{-1} x + (\cos^{-1} x + \sin^{-1} x))$

$= \cos(\cos^{-1} x + \pi/2)$

$= -\sin\cos^{-1} x$ ...... (1)

Let $\cos^{-1} x = \theta$

$\Rightarrow x = \cos\theta$

$\sin\theta = \sqrt{(1-x^2)}$

$\cos^{-1} x = \theta - \sin^{-1}\sqrt{(1-x^2)}$

Now, (1) becomes

$\cos(2\cos^{-1} x + \sin^{-1} x) = -\sin\cos^{-1} x = -\sin^{-1}\sqrt{(1-x^2)}$

$= -\sqrt{(1-x^2)}$ as $-1 \le \sqrt{(1-x^2)} \le 1$

$= -\sqrt{(1-1/25)}$ at $x = 1/5$

$= -2/5\sqrt{6}.$

8. If $\sin^{-1} x + (\sin^{-1} y + \sin^{-1} z) = \cap/2$

Solution:

Find out $x^2 + y^2 + z^2 + 2xyz$.

Let $\sin^{-1} x = A$, $\sin^{-1} y = B$ and $\sin^{-1} z = C$

$\Rightarrow A + B + C = \cap/2$

$A + B = \cap/2 - C.$

Let's operate cos on both sides.

Note: That we have chosen to operate cos because it becomes easier to solve.

$\cos (A + B0 = \cos (\cap/2 - C)$

$\cos A \cos B - \sin A \sin B = \sin C$

$\sqrt{((1-x^2))} \sqrt{((1-y^2))}-xy=z$

$xy + z = \sqrt{((1-x^2)(1-y^2))}$

Squaring both sides

$x^2 y^2 + z^2 + 2xyz = 1 + x^2 y^2 - x^2 - y^2$

$x^2 + y^2 + z^2 = 2xyz = 1$

9. Show that the function $y = 2 \tan^{-1} x + \sin^{-1} 2x/(1-x^2)$ is a constant for $x \geq 1$. Find the value of this constant.

Solution:

We know, $\sin^{-1} 2x/(1+x^2) = 2 \tan^{-1} x$ if $|x| < 1$

Here given $x \geq 1$. So if we convert in above condition then we can get solution.

$x \geq 1$ can be written as $1/x < 1$.

So, for $1/x < 1$

$\sin^{-1} 2x/(1+x^2) = \sin^{-1} (2/x)/(1/x^2 +1) = \sin^{-1} 2(1/x)/((1/x)^2+1) = 2 \tan^{-1}(1/x)$

Now, $y = 2 \tan^{-1} x + 2 \tan^{-1} (1/x)$

$= 2 [\tan^{-1} x + \cot{-1} x]$

$= \cap =$ const.

10. Show that $2\tan^{-1} 1/2 + \tan^{-1} 1/7 = \tan^{-1} 31/17$.

Solution:

We have $2\tan^{-1} 1/2 = \tan^{-1} (2-1/2)/(1-1/4) = \tan^{-1} 4/3$

$\Rightarrow 2\tan^{-1} ((4/3+1/7)/(1-4/21)) = \tan^{-1} 31/17$

11. Solve $2\tan^{-1}(\cos x) = \tan^{-1}(2\operatorname{cosec} x)$.

SS**Solution:**

$= \tan^{-1}[2\cot x \cos ecx] = \tan^{-1}(2 \operatorname{cosec} x)$ (given)

$\Rightarrow \cot x = 1 \Rightarrow x = n\cap + \cap/4.$

12. Show that $\tan^{-1}(x/\sqrt{(a^2-x^2)}) = \sin^{-1}(x/a)$, $a > 0$.

Solution:

Let $\sin^{-1} x/a = \theta$, then $x = a \sin\theta$, $-\pi/2 < \theta < \pi/2$ as $(x \neq a)$
$\Rightarrow x/\sqrt{(a^2-x^2)} = (a \sin\theta)/\sqrt{(a^2-a^2 \sin^2\theta)} = \tan\theta \Rightarrow \tan^{-1}(x/\sqrt{(a^2-x^2)})$
$= \tan^{-1}(\tan\theta) = \theta = \sin^{-1}(x/a)$.

Illustration:

Show that $\tan^{-1}\left(\frac{\cos x}{1+\sin x}\right) = \frac{\pi}{4} - \frac{x}{2}, -\frac{\pi}{2} < x < \frac{3\pi}{2}$.

Solution:

We have $\frac{\cos x}{1+\sin x} = \frac{\sin\left(\frac{\pi}{2}-x\right)}{1+\cos\left(\frac{\pi}{2}-x\right)} = \frac{2\sin\left(\frac{\pi}{2}-\frac{x}{2}\right)\cos\left(\frac{\pi}{4}-\frac{x}{2}\right)}{2\cos^2\left(\frac{\pi}{2}-\frac{x}{2}\right)}$

$= \tan\left(\frac{\pi}{4}-\frac{x}{2}\right) \Rightarrow \tan^{-1}\left(\frac{\cos x}{1+\sin x}\right) = \frac{\pi}{4} - \frac{x}{2}$, as $-\frac{\pi}{2} < \frac{\pi}{4} - \frac{x}{2} < \frac{\pi}{2}$

or $-\frac{\pi}{2} < \frac{x}{2} - \frac{\pi}{4} < \frac{\pi}{2}$ or $\frac{\pi}{2} < x < \frac{3\pi}{2}$.

4

Trigonometric Identities

Trigonometric identities are the equalities involving trigonometric functions and hold true for every value of the variables involved, such that both sides of the equality are defined. In this mini-lesson, we will explore trigonometric identities. There are three primary trigonometric ratios sin, cos, and tan.

]The three other trigonometric ratios sec, cosec, and cot in trigonometry are the reciprocals of sin, cos, and tan respectively. How are these trigonometric ratios (sin, cos, tan, sec, cosec, and cot) connected with each other? They are connected through trigonometric identities (or in short trig identities). Let's understand the trigonometric identities in detail in the following sections.

WHAT ARE TRIGONOMETRIC IDENTITIES?

Trigonometric identities are equations that relate to different trigonometric functions and are true for any value of the variable that is there in the domain. Basically, an identity is an equation that holds true for all the values of the variable(s) present in it.

For example, some of the algebraic identities are:

$(a + b)^2 = a^2 + 2ab + b^2$

$(a - b)^2 = a^2 - 2ab + b^2$

$(a + b)(a-b) = a^2 - b^2$

The algebraic identities relate just the variables whereas the trig identities relate the 6 trigonometric functions sine, cosine, tangent, cosecant, secant, and cotangent. Let's learn about each type of trigonometric identities in detail.

Reciprocal Trigonometric Identities

We already know that the reciprocals of sin, cosine, and tangent are cosecant, secant, and cotangent respectively.

Thus, the reciprocal identities are given as,

- $\sin\theta = 1/\text{cosec}\theta$ (OR) $\text{cosec}\,\theta = 1/\sin\theta$
- $\cos\theta = 1/\sec\theta$ (OR) $\sec\theta = 1/\cos\theta$
- $\tan\theta = 1/\cot\theta$ (OR) $\cot\theta = 1/\tan\theta$

Pythagorean Trigonometric Identities

The Pythagorean trigonometric identities in trigonometry are derived from the Pythagoras theorem. Applying Pythagoras theorem to the right-angled triangle below, we get:

$\text{Opposite}^2 + \text{Adjacent}^2 = \text{Hypotenuse}^2$

Dividing both sides by Hypotenuse^2

$\text{Opposite}^2/\text{Hypotenuse}^2 + \text{Adjacent}^2/\text{Hypotenuse}^2 = \text{Hypotenuse}^2/\text{Hypotenuse}^2$

- $\sin^2\theta + \cos^2\theta = 1$

This is one of the Pythagorean identities. In the same way, we can derive two other Pythagorean trigonometric identities.

- $1 + \tan^2\theta = \sec^2\theta$
- $1 + \cot^2\theta = \text{cosec}^2\theta$

Reciprocal identities

$$\sin u = \frac{1}{\csc u} \quad \cos u = \frac{1}{\sec u} \quad \tan u = \frac{1}{\cot u}$$

$$\csc u = \frac{1}{\sin u} \quad \sec u = \frac{1}{\cos u} \quad \cot u = \frac{1}{\tan u}$$

Pythagorean Identities

$$\sin^2 u + \cos^2 u = 1 \quad 1 + \tan^2 u = \sec^2 u \quad 1 + \cot^2 u = \csc^2 u$$

Quotient Identities

$$\tan u = \frac{\sin u}{\cos u} \quad \cot u = \frac{\cos u}{\sin u}$$

Co-Function Identities

$$\sin\left(\frac{\pi}{2}-u\right)=\cos u \quad \cos\left(\frac{\pi}{2}-u\right)=\sin u \quad \tan\left(\frac{\pi}{2}-u\right)=\cot u$$

$$\csc\left(\frac{\pi}{2}-u\right)=\sec u \quad \sec\left(\frac{\pi}{2}-u\right)=\csc u \quad \cot\left(\frac{\pi}{2}-u\right)=\tan u$$

Even-Odd Identities

$$\sin(-u) = -\sin u \quad \cos(-u) = \cos u \quad \tan(-u) = -\tan u$$

$$\csc(-u) = -\csc u \quad \sec(-u) = \sec u \quad \cot(-u) = -\cot u$$

Sum-Difference Formulas

$$\sin(u \pm v) = \sin u \cos v \pm \cos u \sin v$$
$$\cos(u \pm v) = \cos u \cos v \mp \sin u \sin v$$
$$\tan(u \pm v) = \frac{\tan u \pm \tan v}{1 \mp \tan u \tan v}$$

Double Angle Formulas

$$\begin{aligned} \sin(2u) &= 2\sin u \cos u \\ \cos(2u) &= \cos^2 u - \sin^2 u \\ &= 2\cos^2 u - 1 \\ &= 1 - 2\sin^2 u \\ \tan(2u) &= \frac{2\tan u}{1-\tan^2 u} \end{aligned}$$

Power-Reducing/Half Angle Formulas

$$\sin^2 u = \frac{1-\cos(2u)}{2}$$

$$\cos^2 u = \frac{1+\cos(2u)}{2}$$

$$\tan^2 u = \frac{1-\cos(2u)}{1+\cos(2u)}$$

Sum-to-Product Formulas

$$\sin u + \sin v = 2\sin\left(\frac{u+v}{2}\right)\cos\left(\frac{u-v}{2}\right)$$

$$\sin u - \sin v = 2\cos\left(\frac{u+v}{2}\right)\sin\left(\frac{u-v}{2}\right)$$

$$\cos u + \cos v = 2\cos\left(\frac{u+v}{2}\right)\cos\left(\frac{u-v}{2}\right)$$

$$\cos u - \cos v = -2\sin\left(\frac{u+v}{2}\right)\sin\left(\frac{u-v}{2}\right)$$

Product-to-Sum Formulas

$$\sin u \sin v = \frac{1}{2}\left[\cos(u-v) - \cos(u+v)\right]$$

$$\cos u \cos v = \frac{1}{2}\left[\cos(u-v) + \cos(u+v)\right]$$

$$\sin u \cos v = \frac{1}{2}\left[\sin(u+v) + \sin(u-v)\right]$$

$$\cos u \sin v = \frac{1}{2}\left[\sin(u+v) - \sin(u-v)\right]$$

COMMON SUBSTITUTIONS

1.

$$\int F(ax+b)dx = \frac{1}{a}\int F(u)du$$

where $u = ax + b$

2.

$$\int F\left(\sqrt{ax+b}\right) dx = \frac{2}{a}\int u F(u)\, du$$

where

$$u = \sqrt{ax + b}$$

3.

$$\int F\left(\sqrt[n]{ax+b}\right)\, dx = \frac{n}{a}\int u^{n-1}\, F(u)\, du$$

where

$$u = \sqrt[n]{ax+b}$$

4.

$$\int F\left(\sqrt{a^2 - x^2}\right)\, dx = a\int F(a\cos u)\,\cos u\, du$$

where

$$x = a\sin u$$

5.

$$\int F\left(\sqrt{x^2 + a^2}\right)\, dx = a\int F(a\sec u)\sec^2 u\, du$$

where

$$x = a\tan u$$

6.

$$\int F\left(\sqrt{x^2 - a^2}\right)\, dx = a\int F(a\tan u)\sec u\tan u\, du$$

where

$$x = a\sec u$$

7.

$$\int F(e^{ax})\, dx = \frac{1}{a}\int \frac{F(u)}{u}\, du$$

where

$$u = e^{ax}$$

8.

$$\int F(\ln x)\,dx = \int F(u)\,e^u\,du$$

where

$$u = \ln x$$

9.

$$\int F\left(\sin^{-1}\frac{x}{a}\right)dx = a\int F(u)\cos u\,du$$

where

$$u = \sin^{-1}\frac{x}{a}$$

INTEGRALS CONTAINING ax+b

1.

$$\int \frac{dx}{ax+b} = \frac{1}{a}\ln(ax+b)$$

2.

$$\int \frac{x\,dx}{ax+b} = \frac{x}{a} - \frac{b}{a^2}\ln(ax+b)$$

3.

$$\int \frac{x^3\,dx}{ax+b} = \frac{(ax+b)^3}{3a^4} - \frac{3b(ax+b)^2}{2a^4} + \frac{3b^2(ax+b)}{a^4} - \frac{b^3}{a^4}\ln(ax+b)$$

4.

$$\int \frac{x^2\,dx}{ax+b} = \frac{(ax+b)^2}{2a^3} - \frac{2b(ax+b)}{a^3} + \frac{b^2}{a^3}\ln(ax+b)$$

5.

$$\int \frac{dx}{x(ax+b)} = \frac{1}{b}\ln\left(\frac{x}{ax+b}\right)$$

6.

$$\int \frac{dx}{x^2(ax+b)} = -\frac{1}{bx} + \frac{a}{b^2}\ln\left(\frac{ax+b}{x}\right)$$

7.

$$\int \frac{dx}{x^3(ax+b)} = \frac{2ax-b}{2b^2x^2} + \frac{a^2}{b^3}\ln\left(\frac{x}{ax+b}\right)$$

8.

$$\int \frac{dx}{(ax+b)^2} = \frac{-1}{a(ax+b)}$$

9.

$$\int \frac{x\,dx}{(ax+b)^2} = \frac{b}{a^2(ax+b)} + \frac{1}{a^2}\ln(ax+b)$$

10.

$$\int \frac{x^2\,dx}{(ax+b)^2} = \frac{ax+b}{a^3} - \frac{b^2}{a^3(ax+b)} - \frac{2b}{a^3}\ln(ax+b)$$

11.

$$\int \frac{x^3\,dx}{(ax+b)^2} = \frac{(ax+b)^2}{2a^4} - \frac{3b(ax+b)}{a^4} + \frac{b^3}{a^4(ax+b)} + \frac{3b^2}{a^4}\ln(ax+b)$$

12.

$$\int \frac{dx}{x(ax+b)} = \frac{1}{b(ax+b)} + \frac{1}{b^2}\ln\left(\frac{x}{ax+b}\right)$$

13.

$$\int \frac{dx}{x^2(ax+b)^2} = \frac{-a}{b^2(ax+b)} - \frac{1}{b^2 x} + \frac{2a}{b^3} \ln\left(\frac{ax+b}{x}\right)$$

14.

$$\int \frac{dx}{x^3(ax+b)^2} = -\frac{(ax+b)^2}{2b^4x^2} + \frac{3a(ax+b)}{b^4x} - \frac{a^3x}{b^4(ax+b)} - \frac{3a^2}{b^4} \ln\left(\frac{ax+b}{x}\right)$$

15.

$$\int \frac{dx}{(ax+b)^3} = \frac{-1}{2a(ax+b)^2}$$

16.

$$\int \frac{x\,dx}{(ax+b)^3} = \frac{-1}{a^2(ax+b)} + \frac{b}{2a^2(ax+b)^2}$$

17.

$$\int \frac{x^2\,dx}{(ax+b)^3} = \frac{2b}{a^3(ax+b)} - \frac{b^2}{2a^3(ax+b)^2} + \frac{1}{a^3}\ln(ax+b)$$

18.

$$\int \frac{x^3\,dx}{(ax+b)^3} = \frac{x}{a^3} - \frac{3b^2}{a^4(ax+b)} + \frac{b^3}{2a^4(ax+b)^2} - \frac{3b}{a^4}\ln(ax+b)$$

19.

$$\int \frac{dx}{x(ax+b)^3} = \frac{a^2x^2}{2b^3(ax+b)^2} - \frac{2ax}{b^3(ax+b)} - \frac{1}{b^3}\ln\left(\frac{ax+b}{x}\right)$$

20.

$$\int \frac{dx}{x^2(ax+b)^3} = \frac{-a}{2b^2(ax+b)^2} - \frac{2a}{b^3(ax+b)} - \frac{1}{b^3x} + \frac{3a}{b^4}\ln\left(\frac{ax+b}{x}\right)$$

21.

$$\int \frac{dx}{x^3(ax+b)^3} = \frac{a^4x^2}{2b^5(ax+b)^2} - \frac{4a^3x}{b^5(ax+b)} - \frac{(ax+b)^2}{2b^5x^2} - \frac{6a^2}{b^5}\ln\left(\frac{ax+b}{x}\right)$$

22.

$$\int (ax+b)^n \, dx = \frac{(ax+b)^{n+1}}{(n+1)a}, \quad n \neq -1$$

23.

$$\int x(ax+b)^n \, dx = \frac{(ax+b)^{n+2}}{(n+2)a^2} - \frac{b(ax+b)^{n+1}}{(n+1)a^2}, \quad n \neq -1, -2$$

24.

$$\int x^2(ax+b)^n \, dx = \frac{(ax+b)^{n+3}}{(n+3)a^3} - \frac{2b(ax+b)^{n+2}}{(n+2)a^3} + \frac{b^2(ax+b)^{n+1}}{(n+1)a^3}, \quad n \neq -1, -2, -3$$

25.

$$\int x^m(ax+b)^n \, dx = \begin{cases} \dfrac{x^{m+1}(ax+b)^n}{m+n+1} + \dfrac{nb}{m+n+1}\displaystyle\int x^m(ax+b)^{n-1}\,dx \\[2ex] \dfrac{x^m(ax+b)^{n+1}}{(m+n+1)a} - \dfrac{mb}{(m+n+1)a}\displaystyle\int x^{m-1}(ax+b)^n\,dx \\[2ex] \dfrac{-x^{m+1}(ax+b)^{n+1}}{(n+1)b} + \dfrac{m+n+2}{(n+1)b}\displaystyle\int x^m(ax+b)^{n+1}\,dx \end{cases}$$

INTEGRALS CONTAINING THE SQUARE ROOT OF ax+b

1.

$$\int \frac{dx}{\sqrt{ax+b}} = \frac{2\sqrt{ax+b}}{a}$$

2.

$$\int \frac{x\,dx}{\sqrt{ax+b}} = \frac{2(ax-2b)}{3a^2}\sqrt{ax+b}$$

3.

$$\int \frac{x^2\,dx}{\sqrt{ax+b}} = \frac{2(3a^2x^2-4abx+8b^2)}{15a^3}\sqrt{ax+b}$$

4.

$$\int \frac{dx}{x\sqrt{ax+b}} = \begin{cases} \dfrac{1}{\sqrt{b}}\ln\left(\dfrac{\sqrt{ax+b}-\sqrt{b}}{\sqrt{ax+b}+\sqrt{b}}\right) \\ \\ \dfrac{2}{\sqrt{-b}}\tan^{-1}\sqrt{\dfrac{ax+b}{-b}} \end{cases}$$

5.

$$\int \frac{dx}{x^2\sqrt{ax+b}} = -\frac{\sqrt{ax+b}}{bx} - \frac{a}{2b}\int \frac{dx}{x\sqrt{ax+b}}$$

6.

$$\int \sqrt{ax+b}\,dx = \frac{2\sqrt{(ax+b)^3}}{3a}$$

7.

$$\int x\sqrt{ax+b}\,dx = \frac{2(3ax-2b)}{15a^2}\sqrt{(ax+b)^3}$$

8.

$$\int x^2\sqrt{ax+b}\,dx = \frac{2(15a^2x^2 - 12abx + 8b^2)}{105a^3}\sqrt{(a+bx)^3}$$

9.

$$\int \frac{\sqrt{ax+b}}{x}\,dx = 2\sqrt{ax+b} + b\int \frac{dx}{x\sqrt{ax+b}}$$

10.

$$\int \frac{\sqrt{ax+b}}{x^2}\,dx = -\frac{\sqrt{ax+b}}{x} + \frac{a}{2}\int \frac{dx}{x\sqrt{ax+b}}$$

11.

$$\int \frac{x^m}{\sqrt{ax+b}}\,dx = \frac{2x^m\sqrt{ax+b}}{(2m+1)a} - \frac{2mb}{(2m+1)a}\int \frac{x^{m-1}}{\sqrt{ax+b}}\,dx$$

12.

$$\int \frac{dx}{x^m\sqrt{ax+b}} = -\frac{\sqrt{ax+b}}{(m-1)bx^{m-1}} - \frac{(2m-3)a}{(2m-2)b}\int \frac{dx}{x^{m-1}\sqrt{ax+b}}$$

13.

$$\int x^m\sqrt{ax+b}\,dx = \frac{2x^m}{(2m+3)a}(ax+b)^{3/2} - \frac{2mb}{(2m+3)a}\int x^{m-1}\sqrt{ax+b}\,dx$$

14.

$$\int \frac{\sqrt{ax+b}}{x^m}\,dx = -\frac{\sqrt{ax+b}}{(m-1)x^{m-1}} + \frac{a}{2(m-1)}\int \frac{dx}{x^{m-1}\sqrt{ax+b}}$$

15.

$$\int \frac{\sqrt{ax+b}}{x^m}\,dx - \frac{-(ax+b)^{3/2}}{(m-1)bx^{m-1}} - \frac{(2m-5)a}{(2m-2)b}\int \frac{\sqrt{ax+b}}{x^{m-1}}\,dx$$

16.

$$\int (ax+b)^{m/2}\,dx = \frac{2(ax+b)^{(m+2)/2}}{a(m+2)}$$

17.

$$\int x(ax+b)^{m/2}\,dx = \frac{2(ax+b)^{(m+4)/2}}{a^2(m+4)} - \frac{2b(ax+b)^{(m+2)/2}}{a^2(m+2)}$$

18.

$$\int x^2(ax+b)^{m/2}\,dx = \frac{2(ax+b)^{(m+6)/2}}{a^3(m+6)} - \frac{4b(ax+b)^{(m+4)/2}}{a^3(m+4)} + \frac{2b^2(ax+b)^{(m+2)/2}}{a^3(m+2)}$$

19.

$$\int \frac{(ax+b)^{m/2}}{x}\,dx = \frac{2(ax+b)^{m/2}}{m} + b\int \frac{(ax+b)^{(m-2)/2}}{x}\,dx$$

20.

$$\int \frac{(ax+b)^{m/2}}{x^2}\,dx = -\frac{(ax+b)^{(m+2)/2}}{bx} + \frac{ma}{2b}\int \frac{(ax+b)^{m/2}}{x}\,dx$$

21.

$$\int \frac{dx}{x(ax+b)^{m/2}} = \frac{2}{(m-2)b(ax+b)^{(m-2)/2}} + \frac{1}{b}\int \frac{dx}{x(ax+b)^{(m-2)/2}}$$

INTEGRALS CONTAINING ax+b AND px+q

1.

$$\int \frac{dx}{(ax+b)(px+q)} = \frac{1}{bp-aq} \ln\left(\frac{px+q}{ax+b}\right)$$

2.

$$\int \frac{x\,dx}{(ax+b)(px+q)} = \frac{1}{bp-aq}\left\{\frac{b}{a}\ln(ax+b) - \frac{q}{p}\ln(px+q)\right\}$$

3.

$$\int \frac{dx}{(ax+b)^2(px+q)} = \frac{1}{bp-aq}\left\{\frac{1}{ax+b} + \frac{p}{bp-aq}\ln\left(\frac{px+q}{ax+b}\right)\right\}$$

4.

$$\int \frac{x\,dx}{(ax+b)^2(px+q)} = \frac{1}{bp-aq}\left\{\frac{q}{bp-aq}\ln\left(\frac{ax+b}{px+q}\right) - \frac{b}{a(ax+b)}\right\}$$

5.

$$\int \frac{x^2\,dx}{(ax+b)^2(px+q)} = \frac{b^2}{(bp-aq)a^2(ax+b)} + \frac{1}{(bp-aq)^2}\left\{\frac{q^2}{p}\ln(px+q)\right.$$

$$\left. + \frac{b(bp-2aq)}{a^2}\ln(ax+b)\right\}$$

6.

$$\int \frac{dx}{(ax+b)^m(px+q)^n} = \frac{-1}{(n-1)(bp-aq)}\left\{\frac{1}{(ax+b)^{m-1}(px+q)^{n-1}}\right.$$

$$\left. + a(m+n-2)\int \frac{dx}{(ax+b)^m(px+q)^{n-1}}\right\}$$

7.

$$\int \frac{ax+b}{px+q}dx = \frac{ax}{p} + \frac{bp-aq}{p^2}\ln(px+q)$$

8.

$$\int \frac{(ax+b)^m}{(px+q)^n}dx = \begin{cases} \frac{-1}{(n-1)(bp-aq)}\left\{\frac{(ax+b)^{m+1}}{(px+q)^{n-1}} + (n-m-2)a\int \frac{(ax+b)^{m-1}}{(px+q)^{n-1}}dx\right\} \\ \frac{-1}{(n-m-1)p}\left\{\frac{(ax+b)^m}{(px+q)^{n-1}} + m(bp-aq)\int \frac{(ax+b)^{m-1}}{(px+q)^n}dx\right\} \\ \frac{-1}{(n-1)p}\left\{\frac{(ax+b)^m}{(px+q)^{n-1}} - ma\int \frac{(ax+b)^{m-1}}{(px+q)^{n-1}}dx\right\} \end{cases}$$

INTEGRALS CONTAINING THE SQUARE ROOT OF ax+b AND px+q

1.

$$\int \frac{px+q}{\sqrt{ax+b}}dx = \frac{2(apx+3aq-2bp)}{3a^2}\sqrt{ax+b}$$

2.

$$\int \frac{dx}{(px+q)\sqrt{ax+b}} = \begin{cases} \frac{1}{\sqrt{bp-aq}\sqrt{p}}\ln\left(\frac{\sqrt{p(ax+b)}-\sqrt{bp-aq}}{\sqrt{p(ax+b)}+\sqrt{bp-aq}}\right) \\ \frac{2}{\sqrt{aq-bp}\sqrt{p}}\tan^{-1}\sqrt{\frac{p(ax+b)}{aq-bp}} \end{cases}$$

3.

$$\int \frac{\sqrt{ax+b}}{px+q}dx = \begin{cases} \frac{2\sqrt{ax+b}}{p} + \frac{\sqrt{bp-aq}}{p\sqrt{p}}\ln\left(\frac{\sqrt{p(ax+b)}-\sqrt{bp-aq}}{\sqrt{p(ax+b)}+\sqrt{bp-aq}}\right) \\ \frac{2\sqrt{ax+b}}{p} - \frac{2\sqrt{aq-bp}}{p\sqrt{p}}\tan^{-1}\sqrt{\frac{p(ax+b)}{aq-bp}} \end{cases}$$

4.

$$\int (px+q)^n \sqrt{ax+b}\,dx = \frac{2(px+q)^{n+1}\sqrt{ax+b}}{(2n+3)p} + \frac{bp-aq}{(2n+3)p}\int \frac{(px+q)^n}{\sqrt{ax+b}}\,dx$$

5.

$$\int \frac{(px+q)^n}{\sqrt{ax+b}}\,dx = \frac{2(px+q)^n\sqrt{ax+b}}{(2n+1)a} + \frac{2n(aq-bp)}{(2n+1)a}\int \frac{(px+q)^{n-1}\,dx}{\sqrt{ax+b}}$$

6.

$$\int \frac{\sqrt{ax+b}}{(px+q)^n}\,dx = \frac{-\sqrt{ax+b}}{(n-1)p(px+q)^{n-1}} + \frac{a}{2(n-1)p}\int \frac{dx}{(px+q)^{n-1}\sqrt{ax+b}}$$

7.

$$\int \frac{dx}{(px+q)^n\sqrt{ax+b}} = \frac{\sqrt{ax+b}}{(n-1)(aq-bp)(px+q)^{n-1}}$$

$$+\frac{(2n-3)a}{2(n-1)(aq-bp)}\int \frac{dx}{(px+q)^{n-1}\sqrt{ax+b}}$$

INTEGRALS CONTAINING THE SQUARE ROOTS OF BOTH ax+b AND px+q

1.

$$\int \frac{dx}{\sqrt{(ax+b)(px+q)}} = \begin{cases} \dfrac{2}{\sqrt{ap}}\ln\left(\sqrt{a(px+q)}+\sqrt{p(ax+b)}\right) \\[2ex] \dfrac{2}{\sqrt{-ap}}\tan^{-1}\sqrt{\dfrac{-p(ax+b)}{a(px+q)}} \end{cases}$$

2.

$$\int \frac{x\,dx}{\sqrt{(ax+b)(px+q)}} = \frac{\sqrt{(ax+b)(px+q)}}{ap} - \frac{bp+aq}{2ap}\int \frac{dx}{\sqrt{(ax+b)(px+q)}}$$

3.

$$\int \sqrt{(ax+b)(px+q)}\,dx = \frac{2apx+bp+aq}{4ap}\sqrt{(ax+b)(px+q)} - \frac{(bp-aq)^2}{8ap}\int \frac{dx}{\sqrt{(ax+b)(px+q)}}$$

4.

$$\int \sqrt{\frac{px+q}{ax+b}}\,dx = \frac{\sqrt{(ax+b)(px+q)}}{a} + \frac{aq-bp}{2a}\int \frac{dx}{\sqrt{(ax+b)(px+q)}}$$

5.

$$\int \frac{dx}{(px+q)\sqrt{(ax+b)(px+q)}} = \frac{2\sqrt{ax+b}}{(aq-bp)\sqrt{px+q}}$$

INTEGRALS CONTAINING x^2+a^2

1.

$$\int \frac{dx}{x^2+a^2} = \frac{1}{a}\tan^{-1}\frac{x}{a}$$

2.

$$\int \frac{x\,dx}{x^2+a^2} = \frac{1}{2}\ln(x^2+a^2)$$

3.

$$\int \frac{x^2\,dx}{x^2+a^2} = x - a\tan^{-1}\frac{x}{a}$$

4.

$$\int \frac{x^3\,dx}{x^2+a^2} = \frac{x^2}{2} - \frac{a^2}{2}\ln(x^2+a^2)$$

5.

$$\int \frac{dx}{x(x^2+a^2)} = \frac{1}{2a^2}\ln\left(\frac{x^2}{x^2+a^2}\right)$$

6.

$$\int \frac{dx}{x^2(x^2+a^2)} = -\frac{1}{a^2x} - \frac{1}{a^3}\tan^{-1}\frac{x}{a}$$

7.

$$\int \frac{dx}{x^3(x^2+a^2)} = -\frac{1}{2a^2x^2} - \frac{1}{2a^4}\ln\left(\frac{x^2}{x^2+a^2}\right)$$

8.

$$\int \frac{dx}{(x^2+a^2)^2} = \frac{x}{2a^2(x^2+a^2)} + \frac{1}{2a^3}\tan^{-1}\frac{x}{a}$$

9.

$$\int \frac{x\,dx}{(x^2+a^2)^2} = \frac{-1}{2(x^2+a^2)}$$

10.

$$\int \frac{x^2\,dx}{(x^2+a^2)^2} = \frac{-x}{2(x^2+a^2)} + \frac{1}{2a}\tan^{-1}\frac{x}{a}$$

11.

$$\int \frac{x^3\,dx}{(x^2+a^2)^2} = \frac{a^2}{2(x^2+a^2)} + \frac{1}{2}\ln(x^2+a^2)$$

12.

$$\int \frac{dx}{x(x^2+a^2)^2} = \frac{1}{2a^2(x^2+a^2)} + \frac{1}{2a^4}\ln\left(\frac{x^2}{x^2+a^2}\right)$$

13.

$$\int \frac{dx}{x^2(x^2+a^2)^2} = -\frac{1}{a^4 x} - \frac{x}{2a^4(x^2+a^2)} - \frac{3}{2a^5}\tan^{-1}\frac{x}{a}$$

14.

$$\int \frac{dx}{x^3(x^2+a^2)} = -\frac{1}{2a^2x^2} - \frac{1}{2a^4}\ln\left(\frac{x^2}{x^2+a^2}\right)$$

15.

$$\int \frac{dx}{(x^2+a^2)^n} = \frac{x}{2(n-1)a^2(x^2+a^2)^{n-1}} + \frac{2n-3}{(2n-2)a^2}\int \frac{dx}{(x^2+a^2)^{n-1}}$$

16.

$$\int \frac{x\,dx}{(x^2+a^2)^n} = \frac{-1}{2(n-1)(x^2+a^2)^{n-1}}$$

17.

$$\int \frac{dx}{x(x^2+a^2)^n} = \frac{1}{2(n-1)a^2(x^2+a^2)^{n-1}} + \frac{1}{a^2}\int \frac{dx}{x(x^2+a^2)^{n-1}}$$

18.

$$\int \frac{x^m\,dx}{(x^2+a^2)^n} = \int \frac{x^{m-2}\,dx}{(x^2+a^2)^{n-1}} - a^2\int \frac{x^{m-2}\,dx}{(x^2+a^2)^n}$$

19.

$$\int \frac{dx}{x^m(x^2+a^2)^n} = \frac{1}{a^2}\int \frac{dx}{x^m(x^2+a^2)^{n-1}} - \frac{1}{a^2}\int \frac{dx}{x^{m-2}(x^2+a^2)^n}$$

INTEGRALS CONTAINING x^2-a^2

We assume $x^2 > a^2$:

1.

$$\int \frac{dx}{x^2 - a^2} = \frac{1}{2a}\ln\left(\frac{x-a}{x+a}\right)$$

$$-\frac{1}{a} \text{ or } \coth^{-1}\frac{x}{a}$$

2.

$$\int \frac{x\,dx}{x^2 - a^2} = \frac{1}{2}\ln(x^2 - a^2)$$

3.

$$\int \frac{x^2\,dx}{x^2 - a^2} = x + \frac{a}{2}\ln\left(\frac{x-a}{x+a}\right)$$

4.

$$\int \frac{x^3\,dx}{x^2 - a^2} = \frac{x^2}{2} + \frac{a^2}{2}\ln(x^2 - a^2)$$

5.

$$\int \frac{dx}{x(x^2 - a^2)} = \frac{1}{2a^2}\ln\left(\frac{x^2 - a^2}{x^2}\right)$$

6.

$$\int \frac{dx}{x^2(x^2 - a^2)} = \frac{1}{a^2 x} + \frac{1}{2a^3}\ln\left(\frac{x-a}{x+a}\right)$$

7.

$$\int \frac{dx}{x^3(x^2-a^2)} = \frac{1}{2a^2x^2} - \frac{1}{2a^4}\ln\left(\frac{x^2}{x^2-a^2}\right)$$

8.

$$\int \frac{dx}{(x^2-a^2)^2} = \frac{-x}{2a^2(x^2-a^2)} - \frac{1}{4a^3}\ln\left(\frac{x-a}{x+a}\right)$$

9.

$$\int \frac{x\,dx}{(x^2-a^2)^2} = \frac{-1}{2(x^2-a^2)}$$

10.

$$\int \frac{x^2\,dx}{(x^2-a^2)^2} = \frac{-x}{2(x^2-a^2)} + \frac{1}{4a}\ln\left(\frac{x-a}{x+a}\right)$$

11.

$$\int \frac{x^3\,dx}{(x^2-a^2)^2} = \frac{-a^2}{2(x^2-a^2)} + \frac{1}{2}\ln(x^2-a^2)$$

12.

$$\int \frac{dx}{x(x^2-a^2)^2} = \frac{-1}{2a^2(x^2-a^2)} + \frac{1}{2a^4}\ln\left(\frac{x^2}{x^2-a^2}\right)$$

13.

$$\int \frac{dx}{x^2(x^2-a^2)^2} = -\frac{1}{a^4x} - \frac{x}{2a^4(x^2-a^2)} - \frac{3}{4a^5}\ln\left(\frac{x-a}{x+a}\right)$$

14.

$$\int \frac{dx}{x^3(x^2 - a^2)} = \frac{1}{2a^2x^2} - \frac{1}{2a^4}\ln\left(\frac{x^2}{x^2 - a^2}\right)$$

15.

$$\int \frac{dx}{(x^2 - a^2)^n} = \frac{-x}{2(n-1)a^2(x^2 - a^2)^{n-1}} - \frac{2n-3}{(2n-2)a^2}\int \frac{dx}{(x^2 - a^2)^{n-1}}$$

16.

$$\int \frac{x\,dx}{(x^2 - a^2)^n} = \frac{-1}{2(n-1)(x^2 - a^2)^{n-1}}$$

17.

$$\int \frac{dx}{x(x^2 - a^2)^n} = \frac{-1}{2(n-1)a^2(x^2 - a^2)^{n-1}} - \frac{1}{a^2}\int \frac{dx}{x(x^2 - a^2)^{n-1}}$$

18.

$$\int \frac{x^m\,dx}{(x^2 - a^2)^n} = \int \frac{x^{m-2}\,dx}{(x^2 - a^2)^{n-1}} + a^2\int \frac{x^{m-2}\,dx}{(x^2 - a^2)^n}$$

19.

$$\int \frac{dx}{x^m(x^2 - a^2)^n} = \frac{1}{a^2}\int \frac{dx}{x^{m-2}(x^2 - a^2)^n} - \frac{1}{a^2}\int \frac{dx}{x^m(x^2 - a^2)^{n-1}}$$

INTEGRALS CONTAINING a^2-x^2

We assume $x^2<a^2$:

1.

$$\int \frac{dx}{a^2 - x^2} = \frac{1}{2a}\ln\left(\frac{a+x}{a-x}\right) \quad \text{or} \quad \frac{1}{a}\tanh^{-1}\frac{x}{a}$$

2.

$$\int \frac{x\,dx}{a^2 - x^2} = -\frac{1}{2}\ln(a^2 - x^2)$$

3.

$$\int \frac{x^2\,dx}{a^2 - x^2} = -x + \frac{a}{2}\ln\left(\frac{a+x}{a-x}\right)$$

4.

$$\int \frac{x^3\,dx}{a^2 - x^2} = -\frac{x^2}{2} - \frac{a^2}{2}\ln(a^2 - x^2)$$

5.

$$\int \frac{dx}{x(a^2 - x^2)} = \frac{1}{2a^2}\ln\left(\frac{x^2}{a^2 - x^2}\right)$$

6.

$$\int \frac{dx}{x^2(a^2 - x^2)} = -\frac{1}{a^2 x} + \frac{1}{2a^3}\ln\left(\frac{a+x}{a-x}\right)$$

7.

$$\int \frac{dx}{x^3(a^2 - x^2)} = -\frac{1}{2a^2x^2} + \frac{1}{2a^4}\ln\left(\frac{x^2}{a^2 - x^2}\right)$$

8.

$$\int \frac{dx}{(a^2 - x^2)^2} = \frac{x}{2a^2(a^2 - x^2)} + \frac{1}{4a^3}\ln\left(\frac{a+x}{a-x}\right)$$

9.

$$\int \frac{x\,dx}{(a^2 - x^2)^2} = \frac{1}{2(a^2 - x^2)}$$

10.

$$\int \frac{x^2\,dx}{(a^2-x^2)^2} = \frac{x}{2(a^2-x^2)} - \frac{1}{4a}\ln\left(\frac{a+x}{a-x}\right)$$

11.

$$\int \frac{x^3\,dx}{(a^2-x^2)^2} = \frac{a^2}{2(a^2-x^2)} + \frac{1}{2}\ln(a^2-x^2)$$

12.

$$\int \frac{dx}{x(a^2-x^2)^2} = \frac{1}{2a^2(a^2-x^2)} + \frac{1}{2a^4}\ln\left(\frac{x^2}{a^2-x^2}\right)$$

13.

$$\int \frac{dx}{x^2(a^2-x^2)^2} = \frac{-1}{a^4x} + \frac{x}{2a^4(a^2-x^2)} + \frac{3}{4a^5}\ln\left(\frac{a+x}{a-x}\right)$$

14.

$$\int \frac{dx}{x^3(a^2-x^2)} = -\frac{1}{2a^2x^2} + \frac{1}{2a^4}\ln\left(\frac{x^2}{a^2-x^2}\right)$$

15.

$$\int \frac{dx}{(a^2-x^2)^n} = \frac{x}{2(n-1)a^2(a^2-x^2)^{n-1}} + \frac{2n-3}{(2n-2)a^2}\int \frac{dx}{(a^2-x^2)^{n-1}}$$

16.

$$\int \frac{x\,dx}{(a^2-x^2)^n} = \frac{1}{2(n-1)(a^2-x^2)^{n-1}}$$

17.

$$\int \frac{dx}{x(a^2-x^2)^n} = \frac{1}{2(n-1)a^2(a^2-x^2)^{n-1}} + \frac{1}{a^2}\int \frac{dx}{x(a^2-x^2)^{n-1}}$$

18.

$$\int \frac{x^m\,dx}{(a^2-x^2)^n} = a^2 \int \frac{x^{m-2}\,dx}{(a^2-x^2)^n} - \int \frac{x^{m-2}\,dx}{(a^2-x^2)^{n-1}}$$

19.

$$\int \frac{dx}{x^m(a^2-x^2)^n} = \frac{1}{a^2} \int \frac{dx}{x^m(a^2-x^2)^{n-1}} + \frac{1}{a^2} \int \frac{dx}{x^{m-2}(a^2-x^2)^n}$$

INTEGRALS CONTAINING THE SQUARE ROOT OF x^2+a^2

1.

$$\int \frac{dx}{\sqrt{x^2+a^2}} = \ln\left(x+\sqrt{x^2+a^2}\right) \quad \text{or} \quad \sinh^{-1}\frac{x}{a}$$

2.

$$\int \frac{x\,dx}{\sqrt{x^2+a^2}} = \sqrt{x^2+a^2}$$

3.

$$\int \frac{x^2\,dx}{\sqrt{x^2+a^2}} = \frac{x\sqrt{x^2+a^2}}{2} - \frac{a^2}{2}\ln\left(x+\sqrt{x^2+a^2}\right)$$

4.

$$\int \frac{x^3\,dx}{\sqrt{x^2+a^2}} = \frac{(x^2+a^2)^{3/2}}{3} - a^2\sqrt{x^2+a^2}$$

5.

$$\int \frac{dx}{x\sqrt{x^2+a^2}} = -\frac{1}{a}\ln\left(\frac{a+\sqrt{x^2+a^2}}{x}\right)$$

6.

$$\int \frac{dx}{x^2\sqrt{x^2+a^2}} = -\frac{\sqrt{x^2+a^2}}{a^2x}$$

7.

$$\int \frac{dx}{x^3\sqrt{x^2+a^2}} = -\frac{\sqrt{x^2+a^2}}{2a^2x^2} + \frac{1}{2a^3}\ln\left(\frac{a+\sqrt{x^2+a^2}}{x}\right)$$

8.

$$\int \sqrt{x^2+a^2}\,dx = \frac{x\sqrt{x^2+a^2}}{2} + \frac{a^2}{2}\ln\left(x+\sqrt{x^2+a^2}\right)$$

9.

$$\int x\sqrt{x^2+a^2}\,dx = \frac{(x^2+a^2)^{3/2}}{3}$$

10.

$$\int x^2\sqrt{x^2+a^2}\,dx = \frac{x(x^2+a^2)^{3/2}}{4} - \frac{a^2x\sqrt{x^2+a^2}}{8} - \frac{a^4}{8}\ln\left(x+\sqrt{x^2+a^2}\right)$$

11.

$$\int x^3\sqrt{x^2+a^2}\,dx = \frac{(x^2+a^2)^{5/2}}{5} - \frac{a^2(x^2+a^2)^{3/2}}{3}$$

12.

$$\int \frac{\sqrt{x^2+a^2}}{x}\,dx = \sqrt{x^2+a^2} - a\ln\left(\frac{a+\sqrt{x^2+a^2}}{x}\right)$$

13.

$$\int \frac{\sqrt{x^2+a^2}}{x^2}\,dx = -\frac{\sqrt{x^2+a^2}}{x} + \ln\left(x+\sqrt{x^2+a^2}\right)$$

14.

$$\int \frac{\sqrt{x^2+a^2}}{x}\,dx = \sqrt{x^2+a^2} - a\ln\left(\frac{a+\sqrt{x^2+a^2}}{x}\right)$$

15.

$$\int \frac{dx}{(x^2+a^2)^{3/2}} = \frac{x}{a^2\sqrt{x^2+a^2}}$$

16.

$$\int \frac{x\,dx}{(x^2+a^2)^{3/2}} = \frac{-1}{\sqrt{x^2+a^2}}$$

17.

$$\int \frac{x^2\,dx}{(x^2+a^2)^{3/2}} = \frac{-x}{\sqrt{x^2+a^2}} + \ln\left(x+\sqrt{x^2+a^2}\right)$$

18.

$$\int \frac{x^3\,dx}{(x^2+a^2)^{3/2}} = \sqrt{x^2+a^2} + \frac{a^2}{\sqrt{x^2+a^2}}$$

19.

$$\int \frac{dx}{x(x^2+a^2)^{3/2}} = \frac{1}{a^2\sqrt{x^2+a^2}} - \frac{1}{a^3}\ln\left(\frac{a+\sqrt{x^2+a^2}}{x}\right)$$

20.

$$\int \frac{dx}{x^2(x^2+a^2)^{3/2}} = -\frac{\sqrt{x^2+a^2}}{a^4 x} - \frac{x}{a^4\sqrt{x^2+a^2}}$$

21.

$$\int \frac{dx}{x^3(x^2+a^2)^{3/2}} = \frac{-1}{2a^2x^2\sqrt{x^2+a^2}} - \frac{3}{2a^4\sqrt{x^2+a^2}} + \frac{3}{2a^5}\ln\left(\frac{a+\sqrt{x^2+a^2}}{x}\right)$$

22.

$$\int (x^2+a^2)^{3/2}\,dx = \frac{x(x^2+a^2)^{3/2}}{4} + \frac{3a^2x\sqrt{x^2+a^2}}{8} + \frac{3}{8}a^4\ln\left(x+\sqrt{x^2+a^2}\right)$$

23.

$$\int x(x^2+a^2)^{3/2}\,dx = \frac{(x^2+a^2)^{5/2}}{5}$$

24.

$$\int x^2(x^2+a^2)^{3/2}\,dx = \frac{x(x^2+a^2)^{5/2}}{6} - \frac{a^2x(x^2+a^2)^{3/2}}{24} - \frac{a^4x\sqrt{x^2+a^2}}{16} - \frac{a^6}{16}\ln\left(x+\sqrt{x^2+a^2}\right)$$

25.

$$\int x^3(x^2+a^2)^{3/2}\,dx = \frac{(x^2+a^2)^{7/2}}{7} - \frac{a^2(x^2+a^2)^{5/2}}{5}$$

26.

$$\int \frac{(x^2+a^2)^{3/2}}{x}\,dx = \frac{(x^2+a^2)^{3/2}}{3} + a^2\sqrt{x^2+a^2} - a^3\ln\left(\frac{a+\sqrt{x^2+a^2}}{x}\right)$$

27.

$$\int \frac{(x^2+a^2)^{3/2}}{x^2}\,dx = -\frac{(x^2+a^2)^{3/2}}{x} + \frac{3x\sqrt{x^2+a^2}}{2} + \frac{3}{2}a^2 \ln\left(x+\sqrt{x^2+a^2}\right)$$

28.

$$\int \frac{(x^2+a^2)^{3/2}}{x}\,dx = \frac{(x^2+a^2)^{3/2}}{3} + a^2\sqrt{x^2+a^2} - a^3 \ln\left(\frac{a+\sqrt{x^2+a^2}}{x}\right)$$

INTEGRALS CONTAINING THE SQUARE ROOT OF x^2-a^2

1.

$$\int \frac{dx}{\sqrt{x^2-a^2}} = \ln\left(x+\sqrt{x^2-a^2}\right)$$

2.

$$\int \frac{x\,dx}{\sqrt{x^2-a^2}} = \sqrt{x^2-a^2}$$

3.

$$\int \frac{x^2\,dx}{\sqrt{x^2-a^2}} = \frac{x\sqrt{x^2-a^2}}{2} + \frac{a^2}{2}\ln\left(x+\sqrt{x^2-a^2}\right)$$

4.

$$\int \frac{x^3\,dx}{\sqrt{x^2-a^2}} = \frac{(x^2-a^2)^{3/2}}{3} + a^2\sqrt{x^2-a^2}$$

5.

$$\int \frac{dx}{x\sqrt{x^2-a^2}} = \frac{1}{a}\sec^{-1}\left|\frac{x}{a}\right|$$

6.

$$\int \frac{dx}{x^2\sqrt{x^2-a^2}} = \frac{\sqrt{x^2-a^2}}{a^2 x}$$

7.

$$\int \frac{dx}{x^3\sqrt{x^2-a^2}} = \frac{\sqrt{x^2-a^2}}{2a^2x^2} + \frac{1}{2a^3}\sec^{-1}\left|\frac{x}{a}\right|$$

8.

$$\int \sqrt{x^2-a^2}\,dx = \frac{x\sqrt{x^2-a^2}}{2} - \frac{a^2}{2}\ln\left(x+\sqrt{x^2-a^2}\right)$$

9.

$$\int x\sqrt{x^2-a^2}\,dx = \frac{(x^2-a^2)^{3/2}}{3}$$

10.

$$\int x^2\sqrt{x^2-a^2}\,dx = \frac{x(x^2-a^2)^{3/2}}{4} + \frac{a^2x\sqrt{x^2-a^2}}{8} - \frac{a^4}{8}\ln\left(x+\sqrt{x^2-a^2}\right)$$

11.

$$\int x^3\sqrt{x^2-a^2}\,dx = \frac{(x^2-a^2)^{5/2}}{5} + \frac{a^2(x^2-a^2)^{3/2}}{3}$$

12.

$$\int \frac{\sqrt{x^2-a^2}}{x}\,dx = \sqrt{x^2-a^2} - a\sec^{-1}\left|\frac{x}{a}\right|$$

13.

$$\int \frac{\sqrt{x^2 - a^2}}{x^2}\,dx = -\frac{\sqrt{x^2 - a^2}}{x} + \ln\left(x + \sqrt{x^2 - a^2}\right)$$

14.

$$\int \frac{\sqrt{x^2 - a^2}}{x^3}\,dx = -\frac{\sqrt{x^2 - a^2}}{2x^2} + \frac{1}{2a}\sec^{-1}\left|\frac{x}{a}\right|$$

15.

$$\int \frac{dx}{(x^2 - a^2)^{3/2}} = -\frac{x}{a^2\sqrt{x^2 - a^2}}$$

16.

$$\int \frac{x\,dx}{(x^2 - a^2)^{3/2}} = \frac{-1}{\sqrt{x^2 - a^2}}$$

17.

$$\int \frac{x^2\,dx}{(x^2 - a^2)^{3/2}} = -\frac{x}{\sqrt{x^2 - a^2}} + \ln\left(x + \sqrt{x^2 - a^2}\right)$$

18.

$$\int \frac{x^3\,dx}{(x^2 - a^2)^{3/2}} = \sqrt{x^2 - a^2} - \frac{a^2}{\sqrt{x^2 - a^2}}$$

19.

$$\int \frac{dx}{x(x^2 - a^2)^{3/2}} = \frac{-1}{a^2\sqrt{x^2 - a^2}} - \frac{1}{a^3}\sec^{-1}\left|\frac{x}{a}\right|$$

20.

$$\int \frac{dx}{x^2(x^2-a^2)^{3/2}} = -\frac{\sqrt{x^2-a^2}}{a^4 x} - \frac{x}{a^4\sqrt{x^2-a^2}}$$

21.

$$\int \frac{dx}{x^3(x^2-a^2)^{3/2}} = \frac{1}{2a^2x^2\sqrt{x^2-a^2}} - \frac{3}{2a^4\sqrt{x^2-a^2}} - \frac{3}{2a^5}\sec^{-1}\left|\frac{x}{a}\right|$$

22.

$$\int (x^2-a^2)^{3/2}\,dx = \frac{x(x^2-a^2)^{3/2}}{4} - \frac{3a^2x\sqrt{x^2-a^2}}{8} + \frac{3}{8}a^4\ln\left(x+\sqrt{x^2-a^2}\right)$$

23.

$$\int x(x^2-a^2)^{3/2}\,dx = \frac{(x^2-a^2)^{5/2}}{5}$$

24.

$$\int x^2(x^2-a^2)^{3/2}\,dx = \frac{x(x^2-a^2)^{5/2}}{6} + \frac{a^2x(x^2-a^2)^{3/2}}{24} - \frac{a^4x\sqrt{x^2-a^2}}{16} + \frac{a^6}{16}\ln\left(x+\sqrt{x^2-a^2}\right)$$

25.

$$\int x^3(x^2-a^2)^{3/2}\,dx = \frac{(x^2-a^2)^{7/2}}{7} + \frac{a^2(x^2-a^2)^{5/2}}{5}$$

26.

$$\int \frac{(x^2-a^2)^{3/2}}{x}\,dx = \frac{(x^2-a^2)^{3/2}}{3} - a^2\sqrt{x^2-a^2} + a^3\sec^{-1}\left|\frac{x}{a}\right|$$

27.

$$\int \frac{(x^2-a^2)^{3/2}}{x^2}\,dx = -\frac{(x^2-a^2)^{3/2}}{x} + \frac{3x\sqrt{x^2-a^2}}{2} - \frac{3}{2}a^2\ln\left(x+\sqrt{x^2-a^2}\right)$$

28.

$$\int \frac{(x^2-a^2)^{3/2}}{x^3}\,dx = -\frac{(x^2-a^2)^{3/2}}{2x^2} + \frac{3\sqrt{x^2-a^2}}{2} - \frac{3}{2}a\sec^{-1}\left|\frac{x}{a}\right|$$

INTEGRALS CONTAINING THE SQUARE ROOT OF a^2-x^2

1.

$$\int \frac{dx}{\sqrt{a^2-x^2}} = \sin^{-1}\frac{x}{a}$$

2.

$$\int \frac{x\,dx}{\sqrt{a^2-x^2}} = -\sqrt{a^2-x^2}$$

3.

$$\int \frac{x^2\,dx}{\sqrt{a^2-x^2}} = -\frac{x\sqrt{a^2-x^2}}{2} + \frac{a^2}{2}\sin^{-1}\frac{x}{a}$$

4.

$$\int \frac{x^3\,dx}{\sqrt{a^2-x^2}} = \frac{(a^2-x^2)^{3/2}}{3} - a^2\sqrt{a^2-x^2}$$

5.

$$\int \frac{dx}{x\sqrt{a^2-x^2}} = -\frac{1}{a}\ln\left(\frac{a+\sqrt{a^2-x^2}}{x}\right)$$

6.

$$\int \frac{dx}{x^2\sqrt{a^2-x^2}} = -\frac{\sqrt{a^2-x^2}}{a^2x}$$

7.

$$\int \frac{dx}{x^3\sqrt{a^2-x^2}} = -\frac{\sqrt{a^2-x^2}}{2a^2x^2} - \frac{1}{2a^3}\ln\left(\frac{a+\sqrt{a^2-x^2}}{x}\right)$$

8.

$$\int \sqrt{a^2-x^2}\,dx = \frac{x\sqrt{a^2-x^2}}{2} + \frac{a^2}{2}\sin^{-1}\frac{x}{a}$$

9.

$$\int x\sqrt{a^2-x^2}\,dx = -\frac{(a^2-x^2)^{3/2}}{3}$$

10.

$$\int x^2\sqrt{a^2-x^2}dx = -\frac{x(a^2-x^2)^{3/2}}{4} + \frac{a^2x\sqrt{a^2-x^2}}{8} + \frac{a^4}{8}\sin^{-1}\left(\frac{x}{a}\right)$$

11.

$$\int x^3\sqrt{a^2-x^2}\,dx = \frac{(a^2-x^2)^{5/2}}{5} - \frac{a^2(a^2-x^2)^{3/2}}{3}$$

12.

$$\int \frac{\sqrt{a^2-x^2}}{x}\,dx = \sqrt{a^2-x^2} - a\ln\left(\frac{a+\sqrt{a^2-x^2}}{x}\right)$$

13.

$$\int \frac{\sqrt{a^2 - x^2}}{x^2}\, dx = -\frac{\sqrt{a^2 - x^2}}{x} - \sin^{-1} \frac{x}{a}$$

14.

$$\int \frac{\sqrt{a^2 - x^2}}{x}\, dx = \sqrt{a^2 - x^2} - a \ln\left(\frac{a + \sqrt{a^2 - x^2}}{x}\right)$$

15.

$$\int \frac{dx}{(a^2 - x^2)^{3/2}} = \frac{x}{a^2\sqrt{a^2 - x^2}}$$

16.

$$\int \frac{x\, dx}{(a^2 - x^2)^{3/2}} = \frac{1}{\sqrt{a^2 - x^2}}$$

17.

$$\int \frac{x^2\, dx}{(a^2 - x^2)^{3/2}} = \frac{x}{\sqrt{a^2 - x^2}} - \sin^{-1} \frac{x}{a}$$

18.

$$\int \frac{x^3\, dx}{(a^2 - x^2)^{3/2}} = \sqrt{a^2 - x^2} + \frac{a^2}{\sqrt{a^2 - x^2}}$$

19.

$$\int \frac{dx}{x(a^2 - x^2)^{3/2}} = \frac{1}{a^2\sqrt{a^2 - x^2}} - \frac{1}{a^3} \ln\left(\frac{a + \sqrt{a^2 - x^2}}{x}\right)$$

20.

$$\int \frac{dx}{x^2(a^2 - x^2)^{3/2}} = -\frac{\sqrt{a^2 - x^2}}{a^4 x} + \frac{x}{a^4\sqrt{a^2 - x^2}}$$

21.

$$\int \frac{dx}{x^3(a^2 - x^2)^{3/2}} = \frac{-1}{2a^2x^2\sqrt{a^2 - x^2}} + \frac{3}{2a^4\sqrt{a^2 - x^2}} - \frac{3}{2a^5}\ln\left(\frac{a + \sqrt{a^2 - x^2}}{x}\right)$$

22.

$$\int (a^2 - x^2)^{3/2}\, dx = \frac{x(a^2 - x^2)^{3/2}}{4} + \frac{3a^2x\sqrt{a^2 - x^2}}{8} + \frac{3}{8}a^4 \sin^{-1}\frac{x}{a}$$

23.

$$\int x(a^2 - x^2)^{3/2}\, dx = -\frac{(a^2 - x^2)^{3/2}}{5}$$

24.

$$\int x^2(a^2 - x^2)^{3/2}\, dx = -\frac{x(a^2 - x^2)^{5/2}}{6} + \frac{a^2x(a^2 - x^2)^{3/2}}{24}$$

$$+ \frac{a^4x\sqrt{a^2 - x^2}}{16} + \frac{a^6}{16}\sin^{-1}\frac{x}{a}$$

25.

$$\int x^3(a^2 - x^2)^{3/2}\, dx = \frac{(a^2 - x^2)^{7/2}}{7} - \frac{a^2(a^2 - x^2)^{5/2}}{5}$$

26.

$$\int \frac{(a^2 - x^2)^{3/2}}{x} dx = \frac{(a^2 - x^2)^{3/2}}{3} + a^2\sqrt{a^2 - x^2} - a^3 \ln\left(\frac{a + \sqrt{a^2 - x^2}}{x}\right)$$

27.

$$\int \frac{(a^2 - x^2)^{3/2}}{x^2} dx = -\frac{(a^2 - x^2)^{3/2}}{x} - \frac{3x\sqrt{a^2 - x^2}}{2} - \frac{3}{2}a^2 \sin^{-1}\frac{x}{a}$$

28.

$$\int \frac{(a^2 - x^2)^{3/2}}{x} dx = \frac{(a^2 - x^2)^{3/2}}{3} + a^2\sqrt{a^2 - x^2} - a^3 \ln\left(\frac{a + \sqrt{a^2 - x^2}}{x}\right)$$

First and Second Order Differential Equations

First Order Differential equations

A first order differential equation is of the form:

$$\frac{dy}{dx} = f(x, y)$$

Linear Equations:

$$\frac{dy}{dx} + p(x)y = q(x).$$

The general general solution is given by

$$y = \frac{\int u(x)q(x)dx + C}{u(x)};$$

where

$$u(x) = \exp\left(\int p(x)dx\right);$$

is called the integrating factor.

Separable Equations:

$$\frac{dy}{dx} = h(x)g(y)$$

(1)
Solve the equation g(y) = 0 which gives the constant solutions.
(2)
The non-constant solutions are given by

$$\int \frac{1}{g(y)} dy = \int h(x)dx.$$

Bernoulli Equations:

$$\frac{dy}{dx} + p(x)y = q(x)y^n.$$

(1)

Consider the new function . $v = y^{1-n}$

(2)
The new equation satisfied by v is

$$\frac{dv}{dx} + (1-n)p(x)v = (1-n)q(x).$$

(3)
Solve the new linear equation to find v.

(4) $y = v^{1/(1-n)}$

Back to the old function y through the substitution .
(5)
If n > 1, add the solution y=0 to the ones you got in (4).
Homogenous Equations:

$$\frac{dy}{dx} = f(x,y)$$

is homogeneous if the function f(x,y) is homogeneous, that is

$$f(tx,ty) = f(x,y) \text{ for any number } t.$$

By substitution, we consider the new function

$$z = \frac{y}{x} \text{ (which is equivalent to } y = xz\text{)}.$$

The new differential equation satisfied by z is

$$x\frac{dz}{dx} + z = f(1, z)$$

which is a separable equation. The solutions are the constant ones f(1,z) - z =0 and the non-constant ones given by

$$\ln|x| = \int \frac{dz}{f(1,z) - z} + C$$

Do not forget to go back to the old function y = xz.

Exact Equations:

$$M(x, y)dx + N(x, y)dy =$$

is exact if

$$\frac{\partial M}{\partial y} = \frac{\partial N}{\partial x}.$$

The condition of exactness insures the existence of a function F(x,y) such that

$$\begin{cases} \dfrac{\partial F}{\partial x} = M(x, y); \\[2ex] \dfrac{\partial F}{\partial y} = N(x, y). \end{cases}$$

All the solutions are given by the implicit equation

Second Order Differential equations

Homogeneous Linear Equations with constant coefficients:

$$ay'' + by' + cy = 0 \qquad (a \neq 0)$$

Write down the characteristic equation

$$ar^2 + br + c = 0.$$

(1)

If F1and F2 are distinct real numbers (this happens if $b^2 - 4ac > 0$), then the general solution is

$$y = c_1 e^{r_1 x} + c_2 e^{r_2 x}.$$

(2)

If (which happens if), then the general solution is

(3)

If and are complex numbers (which happens if $b^2 - 4ac < 0$), then the general solution is

$$y = c_1 e^{\alpha x} \cos(\beta x) + c_2 e^{\alpha x} \sin(\beta x),$$

where

$$r_1 = \overline{r_2} = \alpha + i\beta = \frac{-b}{2a} + i\frac{\sqrt{4ac - b^2}}{2a}$$

that is

$$\alpha = \frac{-b}{2a} \text{ and } \beta = \frac{\sqrt{4ac - b^2}}{2a}.$$

Non Homogeneous Linear Equations:

$$y'' + p(x)y' + q(x)y = g(x).$$

The general solution is given by

$$y = y_p(x) + y_h(x)$$

where $y_p(x)$ is a particular solution and $y_h(x)$ is the general solution of the associated homogeneous equation

$$y'' + p(x)y' + q(x)y = 0.$$

In order to find two major techniques were developed.

Method of undetermined coefficients or Guessing Method

This method works for the equation

$$ay'' + by' + cy = g(x)$$

where a, b, and c are constant and

$$g(x) = P_n(x)e^{\alpha x}\cos(\beta x), \quad \text{or} \quad g(x) = P_n(x)e^{\alpha x}\sin(\beta x)$$

where $P_n(x)$ is a polynomial function with degree n. In this case, we have

$$y_p(x) = x^s\left(T_n(x)e^{\alpha x}\cos(\beta x) + R_n(x)e^{\alpha x}\sin(\beta x)\right),$$

where

$$T_n(x) = A_0 + A_1 x + \ldots + A_n x^n, \quad \text{and} \quad R_n(x) = B_0 + B_1 x + \ldots + B_n x^n$$

The constants A_i and B_i have to be determined. The power s is equal to 0 if $\alpha + i\beta$ is not a root of the characteristic equation. $\alpha + i\beta$ If is a simple root, then s=1 and s=2 if it is a double root.

Remark. If the nonhomogeneous term g(x) satisfies the following

$$g(x) = \sum_{i=1}^{i=N} g_i(x).$$

where are of the forms cited above, then we split the original equation into N equations

$$ay'' + by' + cy = g_i(x) \quad (i = 1, 2, .., N)$$

then find a particular solution . A particular solution to the original equation is given by

$$y_p(x) = \sum_{i=1}^{i=N} y_i.$$

Method of Variation of Parameters

This method works as long as we know two linearly independent solutions $\{y_1, y_2\}$ of the homogeneous equation

$$a(x)y'' + b(x)y' + c(x)y = 0$$

Note that this method works regardless if the coefficients are constant or not. a particular solution as

$$y_p(x) = u_1(x)y_1(x) + u_2(x)y_2(x).$$

where u_1 and u_2 are functions. From this, the method got its name.

The functions u_1 and u_2 are solutions to the system:

$$\begin{cases} u_1'y_1 + u_2'y_2 = 0 \\ u_1'y_1' + u_2'y_2' = g(x) \end{cases}$$

which implies

$$\begin{cases} u_1(x) = -\displaystyle\int \frac{y_2(x)g(x)}{W(y_1, y_2)(x)}dx; \\ \\ u_2(x) = \displaystyle\int \frac{y_1(x)g(x)}{W(y_1, y_2)(x)}dx\,. \end{cases}$$

Therefore, we have

$$y_p(x) = -y_1(x)\int \frac{y_2(x)g(x)}{W(y_1, y_2)(x)}dx + y_2(x)\int \frac{y_1(x)g(x)}{W(y_1, y_2)(x)}dx.$$

Euler-Cauchy Equations:

$$x^2y'' + bxy' + cy = 0$$

where b and c are constant numbers. By substitution, set

$$x = e^t$$

then the new equation satisfied by y(t) is

$$r^2 + (b-1)r + c = 0\,.$$

which is a second order differential equation with constant coefficients.

(1) $$r^2 + (b-1)r + c = 0\,.$$

Write down the characteristic equation

(2) $$y(x) = c_1|x|^{r_1} + c_2|x|^{r_2}.$$

If the roots r_1 and r_2 are distinct real numbers, then the general solution is given by

(2) $$y(x) = \left(c_1 + c_2 x\right)|x|^{r_1}.$$

If the roots and are equal ( ), then the general solution is

(3)

If the roots r_1 and r_2 are complex numbers, then the general solution is

$$y = \left(c_1 \cos(\beta x) + c_2 \sin(\beta x)\right)|x|^{\alpha}$$

where and . $\alpha = -(b-1)/2 \beta = \sqrt{4c - (b-1)^2}/2$

5

Sine

The sine of an angle is a trigonometric function that is denoted by sin x, where x is the angle in consideration. In a right-angled triangle, the ratio of the perpendicular and the hypotenuse is called the sine function.

In other words, it is the ratio of the side opposite to the angle in consideration and the hypotenuse and its value vary as the angle varies. The sine function is used to represent sound and light waves in the field of physics.

In this article, we will learn the basic properties of sin x, sine graph, its domain and range, derivative, integral, and power series expansion. The sine function is a periodic function and has a period of 2∩. We will solve a few examples using the sine function for a better understanding of the concept.

WHAT IS SINE?

The sine of an angle in a right-angled triangle is a ratio of the side opposite to an angle and the hypotenuse.

The sine function is an important periodic function in trigonometry and has a period of 2Π. To understand the derivation of sin x, let us consider a unit circle centered at the origin of the coordinate plane. A variable point P moves on the boundary (circumference) of this circle.

Observe that P is in the first quadrant, and OP makes an acute angle of x radians with the positive x-axis. PQ is the perpendicular dropped from P onto the horizontal axis. The triangle is thus formed by joining the points O, P, and Q as shown in the figure.

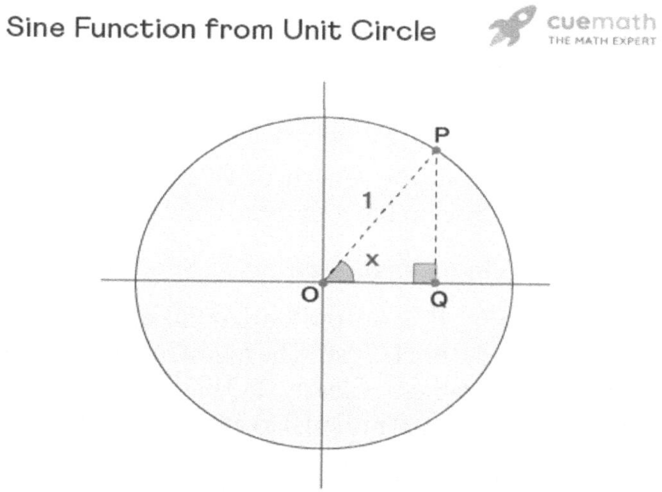

Hence, the sine function for the above case can be mathematically written as sin x = PQ/OP. Here, x is the acute angle between the hypotenuse and the base of a right-angled triangle OPQ.

Sine Definition

The sine of an angle can be defined as the ratio of the length of the perpendicular to the length of the hypotenuse in a right-angled triangle. It gives the value of the sine function of the angle between the base and hypotenuse of the right triangle. Mathematically, it is abbreviated as sin x, where x is an acute angle between the base and hypotenuse of the right triangle.

Sine Function Formula

The sine function is written as the ratio of the length of the perpendicular and hypotenuse of the right-angled triangle. Mathematically, the sine function formula in terms of sides of a right-angled triangle is written as:

sin x = Opposite Side/Hypotenuse = Perpendicular/Hypotenuse

Sine Function Domain and Range

The domain of sine function is all real numbers as sin x is defined for all x in $(-\infty, \infty)$. Whereas the range of sin x is as the value of sin x does not go beyond this. The graph of sine function looks like a wave that oscillates between -1 and 1. Also, the period of sin x is 2Π as its value repeats after every 2Π radians. The domain and range of sine function can also be observed using its graph. In the section below, let us see how the graph of sin x is made.

Sine Graph

Before getting to the graph of the sine function, let us understand how the values of sine vary on a unit circle and then plot them on the graph. As shown in the image above, we note that sin x = PQ/OP = PQ/1 = PQ (As the radius of a unit circle is 1, so OP = 1). As x varies, the value of sin x varies with the variation in the length of PQ. Now, we will study the variation in the sine function in the four quadrants of the coordinate plane.

Case 1: Variation of PQ in the first quadrant.

Suppose that initially, P is on the horizontal axis. Let us consider a movement of P through 90° or Π/2 rad. The following figure shows different positions of P for this movement. Clearly, PQ has increased in length, from an initial value of 0 (when x is 0 radians) to a final value of 1 (when x is Π/2 radians).

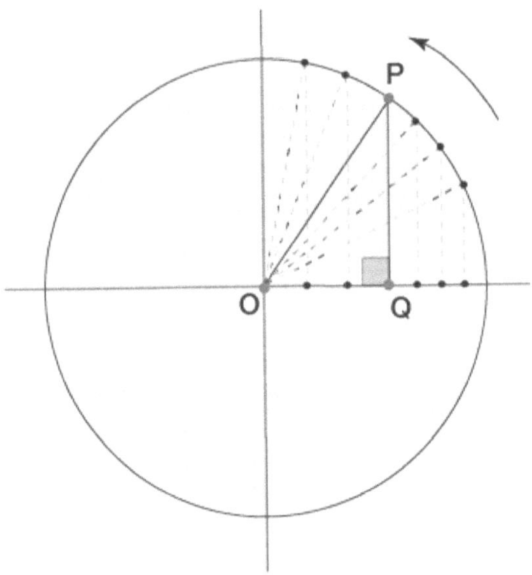

Case 2: ***Variation of PQ in the second quadrant.***

Now, we will check the position of P in the second quadrant as we did in the first quadrant and check how the value of the sine function varies. P subsequently moves from 90° position to 180° position. In this phase of the movement, the length of PQ decreases, from a maximum of 1 at 90°, to a minimum of 0 at 180°.

Case 3: ***Variation of PQ in the third quadrant.***

When P moves from a position of 180° to a position of 270°, though the length or magnitude of PQ increases. But since the direction is along the negative y-axis, the actual value of sine function decreases from 0 to - 1. Thus, the value of the sine function for angle x decreases.

Case 4: ***Variation of PQ in the fourth quadrant.***

Finally, when P moves from a position of 270° to a position of 360°, sin x increases from −1 to 0 (once again). Though the length or magnitude of PQ decreases but the magnitude value of PQ will increase because its direction is along the negative y-axis. Thus, the value of the sine for angle x increases.

We can now plot this variation on a graph. The horizontal axis represents the input variable x as the angle in radians, and the vertical axis represents the value of the sine function. Merging the response of variation in the value of PQ for all four quadrants, we obtained the complete plot of PQ vs x or sin x vs x, for one complete cycle of 0 radians to 2Π radians (0° to 360°). The sine graph thus obtained is shown below:

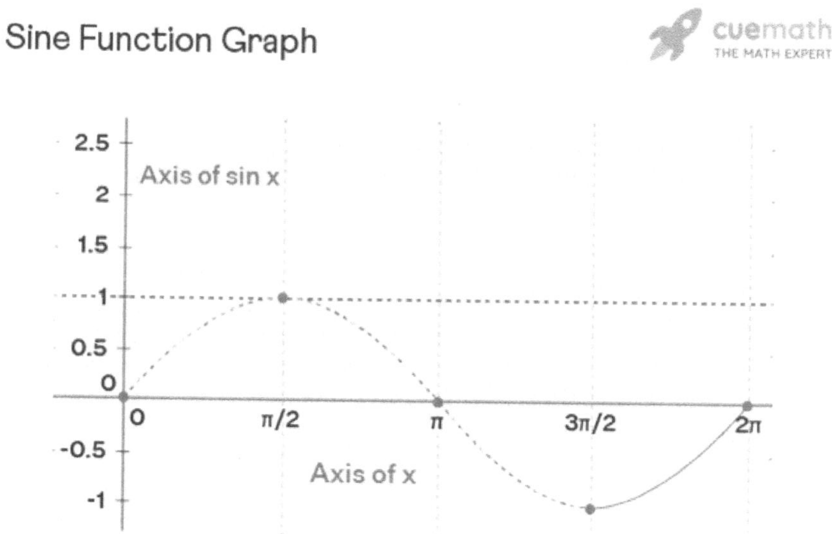

Sine Table

Now, let us go through the values of the sine function for some specific angles such as 0°, 30°, 45°, 60°, 90°, etc. as they are easy to remember. Most of the values given below are used for solving different problems in trigonometry. The values of sin x are listed below in the trigonometric table:

Sine Degrees	Sine Radians	Value of Sine Function (sin x)
sin 0°	sin 0	0
sin 30°	sin Π/6	1/2
sin 45°	sin Π/4	1/√2
sin 60°	sin Π/3	√3/2
sin 90°	sin Π/2	1
sin 120°	sin 2Π/3	√3/2
sin 150°	sin 5Π/6	1/2
sin 180°	sin Π	0
sin 270°	sin 3Π/2	-1
sin 360°	sin 2Π	0

Properties of Sine Function

Properties of sine function depend upon the quadrant in which the angle lies. Sine is a special trigonometric function and has many properties. Some of them are listed below:

- The sine graph repeats itself after a period of 2π, which shows the function is periodic with a period of 2π.
- Power series expansion of the sine function is sin x = Σ∞n=0(−1)nx2n+1(2n+1)Σn=0∞(−1)nx2n+1(2n+1)!
- The sine function is an odd function because sin(−x) = −sin x.
- The reciprocal of the sin x is the cosec x.
- The domain of the sine is all real numbers and the range is.

Sine Identities

In trigonometry, there are several identities involving the sine function. These identities are very useful in solving various math problems. Some of them are listed below:

- sin x = 1/ cosec x
- Inverse of sine is $\sin^{-1}x$ = arcsin x, where x lies in
- $\sin^2 x + \cos^2 x = 1$
- sin (x + y) = sin x cos y + sin y cos x
- sin (x - y) = sin x cos y - sin y cos x
- sin 2x = 2 sin x cos x
- Derivative of sin x: d(sin x)/dx = cos x
- Integral of sin x: ∫sin x dx = -cos x + C, where C is the constant of integration

Inverse Sine

Inverse sine is one of the inverse trigonometric functions and it is written as $\sin^{-1}x$ and is read as "sin inverse x". It is also written as arcsin (x) (read as 'arc sine x'). Please note that $\sin^{-1}x$ (sin inverse x) is not the same as $(\sin x)^{-1}$ (reciprocal of sin x which is csc x). We have 6 inverse trigonometric functions such as:

- arcsin x = $\sin^{-1}x$ = inverse of sin x
- arccos x = $\cos^{-1}x$ = inverse of cos x
- arctan x = $\tan^{-1}x$ = inverse of tan x
- arccsc x = $\csc^{-1}x$ = inverse of csc x
- arcsec x = $\sec^{-1}x$ = inverse of sec x
- arccot x = $\cot^{-1}x$ = inverse of cot x

Here, we will study in detail about the inverse sine function (sin inverse) along with its graph, domain, range, and properties. Also, we will learn the formulas, derivatives, and integral of sin inverse x along with a few solved examples for a better understanding of the concept.

What is Inverse Sine?

The inverse sine function is the inverse of the sine function and thus it is one of the inverse trigonometric functions. It is also known as arcsin function which is pronounced as "arc sin". It is mathematically written as "asin x" (or) "$\sin^{-1}x$" or "arcsin x".

We read "$\sin^{-1}x$" as "sin inverse of x". We know that if two functions f and f^{-1} are inverses of each other, then $f(x) = y \Rightarrow x = f^{-1}(y)$. So $\sin x = y \Rightarrow x = \sin^{-1}(y)$. i.e., when "sin" moves from one side to the other side of the equation, it becomes $\sin^{-1}$. Let us consider a few examples to see how the inverse sine function works.

Inverse Sine Examples

- $\sin 0 = 0 \Rightarrow 0 = \sin^{-1}(0)$
- $\sin \pi 2 = 1 \Rightarrow \pi/2 = \sin^{-1}(1)$
- $\sin \pi/6 = 0.5 \Rightarrow \pi/6 = \sin^{-1}(0.5)$

Sin Inverse x Formula

In a right-angled triangle, the sine of an angle (θ) is the ratio of its opposite side to the hypotenuse. i.e., sin θ = (opposite side) / (hypotenuse). Then by the definition of inverse sine, θ = $\sin^{-1}$[(opposite side) / (hypotenuse)]. As the image below shows, to find the measure of the angle θ, we use the sin inverse x formula given by, θ = arcsin[(opposite side) / (hypotenuse)]. Inverse sine gives the measure of the angle for the corresponding value of the sine function such that the angle lies in the range of the sin inverse (We will discuss the domain and range of inverse sine in the next section.).

Definition of Inverse Sine

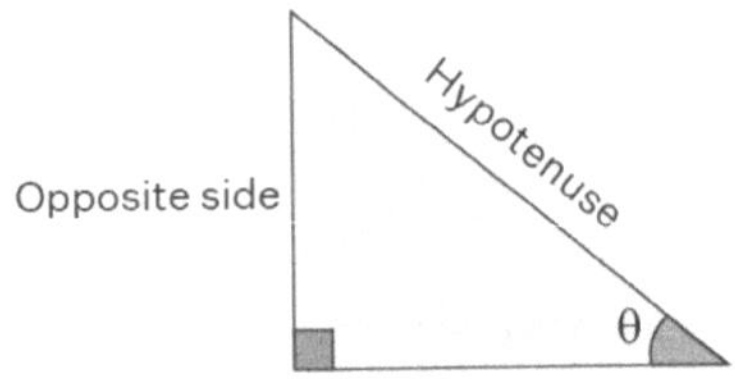

$$\sin\theta = \frac{\text{Opposite side}}{\text{Hypotenuse}}$$

$$\Rightarrow \theta = \sin^{-1}\left(\frac{\text{Opposite side}}{\text{Hypotenuse}}\right)$$

(or)

$$\text{Arcsin}\left(\frac{\text{Opposite side}}{\text{Hypotenuse}}\right)$$

Thus, the inverse sine function is used to find the angle in a right-angled triangle when the opposite side and the hypotenuse are given. Also, it can be used to find the unknown angles in any triangle by using the sine law. In a triangle ABC, if AB = c, BC = a, and CA = b, then by the law of sines,

(sin A) / a = (sin B) / b

From this,

sin A = (a sin B) / b

A = $\sin^{-1}$ [(a sin B) / b]

Likewise, we can find the other angles of the triangle.

Domain and Range of Inverse Sine

In this section, let us see how can we find the domain and range of the inverse sine function. We know that the sine function is a function from R →. But sine function is NOT one-one on the domain R and hence its inverse does not exist. For the sine function to be one-one, its domain can be restricted to one of the intervals [-3Π/2, -Π/2], [-Π/2, Π/2], [Π/2, 3Π/2], etc. Corresponding to each of these intervals, we get a branch of the inverse sine. But the domain of sine function is usually restricted to [-Π/2,Π/2] to make it one-one. The branch of sin inverse when the domain of sine function is [-Π/2, Π/2] is called the principal value branch.

We know that the inverse of a function exists if and only if it is bijective and the domain and range of a function are interchanged to be the range and domain of its inverse function respectively. Hence,

- the domain of sin inverse x is
- the range of sin inverse x is $[-\pi/2, \pi/2]$.
  i.e., arcsin x (or) $\sin^{-1}x : \rightarrow [-\Pi/2, \Pi/2]$

Sin Inverse Graph

The graph of the inverse sine function with its range to be principal branch $[-\Pi/2, \Pi/2]$ can be drawn using the following table. Here, we have chosen random values for x in the domain of sin inverse x which is. We know the values of the sine function using the trigonometric table and using the functioning of the sin inverse function, we have the following table.

x	$y = \sin^{-1}x$
-1	$\sin^{-1}(-1) = -\sin^{-1}(1) = -\Pi/2$
-0.5	$\sin^{-1}(-0.5) = -\sin^{-1}(0.5) = -\Pi/6$
0	$\sin^{-1}(0) = 0$
0.5	$\sin^{-1}(0.5) = \Pi/6$
1	$\sin^{-1}(1) = \Pi/2$

By plotting these points on the graph, we get the sin inverse graph.

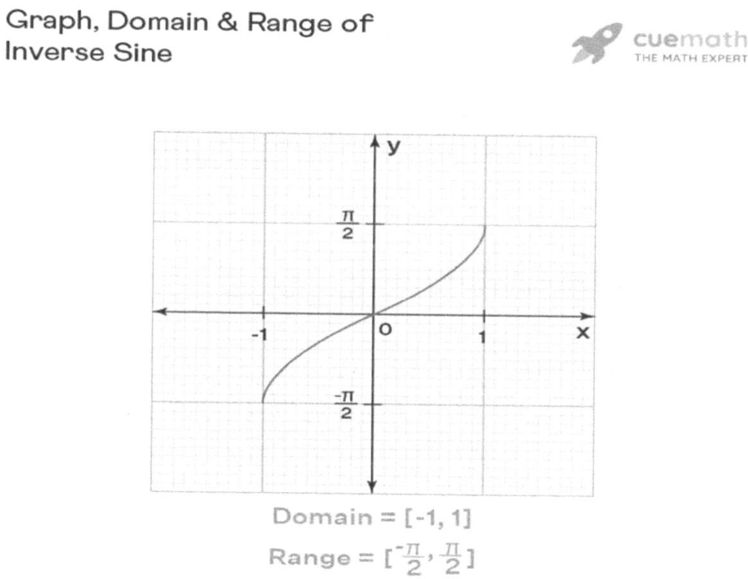

Steps to Find Sin Inverse x

Here are the steps to find the sin inverse of x.

- Since the range of sin inverse x is $[-\pi/2, \pi/2]$, the answer should lie in this interval.

- Assume that $y = \sin^{-1}x$. Then by the definition of inverse sine, $\sin y = x$.
- Think what value of y in the interval $[-\pi/2, \pi/2]$ satisfies the equation $\sin y = x$ and that is the answer.

  Here are some examples to understand these steps.

  Examples of Finding Sin Inverse of x

Note that $\sin^{-1}x$ should always result in some angle that lies in the interval $[-\Pi/2, \Pi/2]$.

- $\sin^{-1}(1) = \pi/2$ as $\sin \pi/2 = 1$
- $\sin^{-1}(-1) = -\pi/2$ as $\sin(-\pi/2) = -1$
- $\sin^{-1}(-0.5) = -\pi/6$ as $\sin(-\pi/6) = -0.5$

Properties of Inverse Sine Function

Here are some properties/formulas of inverse sine. These are very helpful in solving the problems related to sin inverse in trigonometry.

- $\sin(\sin^{-1}x) = x$ only when $x \in [-1, 1]$
  ([When x $\in [-1, 1]$, $\sin(\sin^{-1}x)$ is NOT defined)
- $\sin^{-1}(\sin x) = x$, only when $x \in [-\pi/2, \pi/2]$. To know how to calculate $\sin^{-1}(\sin x)$ when x $\in [-\pi/2, \pi/2]$, click here.
- $\sin^{-1}(-x) = -\sin^{-1}x$, when $x \in [-1, 1]$
- $\sin^{-1}(1/x) = \csc^{-1}x$, when $|x| \geq 1$
- $\sin^{-1}x + \cos^{-1}x = \pi/2$, when $x \in [-1, 1]$
- $\sin^{-1}(2x\sqrt{1 - x^2}) = 2\sin^{-1}x$, when $-1/\sqrt{2} \leq x \leq 1/\sqrt{2}$ and
  $\sin^{-1}(2x\sqrt{1 - x^2}) = 2\cos^{-1}x$, when $1/\sqrt{2} \leq x \leq 1$

Derivative of Inverse Sine

Let us find the derivative of $y = \sin^{-1}x$. By the definition of inverse sine, $y = \sin^{-1}x$ can be written as $\sin y = x$. Differentiating this on both sides with respect to x using the chain rule,

$\cos y\ (dy/dx) = 1$

$dy/dx = 1/\cos y$ ... (1)

Now, we have $\sin^2 y + \cos^2 y = 1 \Rightarrow \cos^2 y = 1 - \sin^2 y \Rightarrow \cos y = \sqrt{1 - \sin^2 y} = \sqrt{1 - x^2}$

Substituting this in (1),

$dy/dx = 1/\sqrt{1 - x^2}$

Thus, the inverse sine derivative (or) the derivative of sin inverse x is $1/\sqrt{1 - x^2}$.

Integral of Inverse Sine

We will find $\int \sin^{-1}x\, dx$ using the integration by parts. For this, we write the above integral as

$\int \sin^{-1}x \cdot 1\, dx$

Using LIATE, $f(x) = \sin^{-1}x$ and $g(x) = 1$.

By integration by parts,

$\int f(x) . g(x)\, dx = f(x) \int g(x)\, dx \in \int (f2\, (x) \int g(x)\, dx)\, dx + C$

$\int \sin^{-1}x \cdot 1\, dx = \sin^{-1}x \int 1\, dx - \int [d/dx(\sin^{-1}x) \int x\, dx] + C$

$\int \sin^{-1}x\, dx = \sin^{-1}x\, (x) - \int [1/"1 - x^2]\, x\, dx + C$

Let us evaluate the integral on the right side using u-substitution method. For this, assume that $1 - x^2 = u$. From this, $-2x\, dx = du$ (or) $x\, dx = -1/2\, du$.

$\int \sin^{-1}x\, dx = x \sin^{-1}x - \int (1/\sqrt{u})\, (-1/2)\, du + C$

$= x \sin^{-1}x + 1/2 \int u^{-1/2}\, du + C$

$= x \sin^{-1}x + (1/2)\, (u^{1/2}/(1/2)) + C$

$= x \sin^{-1}x + \sqrt{u} + C$

$= x \sin^{-1}x + \sqrt{(1 - x^2)} + C$

Therefore, $\int \sin^{-1}x\, dx = x \sin^{-1}x + \sqrt{1 - x^2} + C$.

Important Notes on Sin Inverse:

- Inverse sine can be written as $\sin^{-1}$ (or) arcsin (or) asin and it is a function with domain and range $[-\pi/2, \pi/2]$.
- Inverse sin is NOT same as $(\sin x)^{-1}$ as $(\sin x)^{-1} = 1/(\sin x) = \csc x$.
- $\sin(\sin^{-1}x)$ is NOT always x. $\sin(\sin^{-1}x) = x$ only when $x \in [-1, 1]$.
- $\sin^{-1}(\sin x)$ is NOT always x. $\sin^{-1}(\sin x) = x$ only when $x \in [-\pi/2, \pi/2]$.

Law of Sines

The law of sines establishes the relationship between the sides and angles of an oblique triangle(non-right triangle). Law of sines and law of cosines in trigonometry are important rules used for "solving a triangle". According to the sine rule, the ratios of the side lengths of a triangle to the sine of their respective opposite angles are equal. Let us understand the sine law formula and its proof using solved examples in the following sections.

What is Law of Sines?

The law of sines relates the ratios of side lengths of triangles to their respective opposite angles. This ratio remains equal for all three sides and opposite angles. We can therefore apply the sine rule to find the missing angle or side of any triangle using the requisite known data.

LAW OF SINES: DEFINITION

The ratio of the side and the corresponding angle of a triangle is equal to the diameter of the circumcircle of the triangle. The sine law is can therefore be given as,

a/sinA = b/sinB = c/sinC = 2R

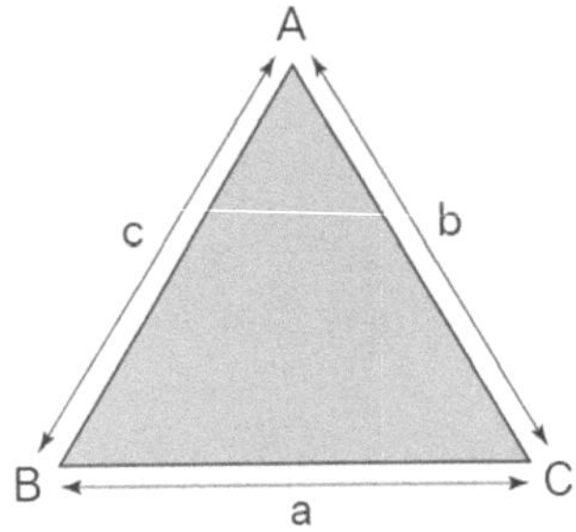

- Here a, b, c are the lengths of the sides of the triangle.
- A, B, and C are the angle of the triangle.
- R is the radius of the circumcircle of the triangle.

Law of Sines Formula

The law of sines formula is used for relating the lengths of the sides of a triangle to the sines of consecutive angles. It is the ratio of the length of the side of the triangle to the sine of the angle thus formed between the other two remaining sides. The law of sines formula is used for any triangle apart from SAS triangle and SSS triangle. It says,

a/sin A = b/sin B = c/sin C

where,

- a, b, and c are the lengths of the triangle
- A, B, and C are the angles of the triangle.

This formula can be represented in three different forms given as,

- a/sinA = b/sinB = c/sinC
- sinA/a = sinB/b = sinC/c
- a/b = sinA/sinB; a/c = sinA/sinC; b/c = sinB/sinC

Law of Sines Formula

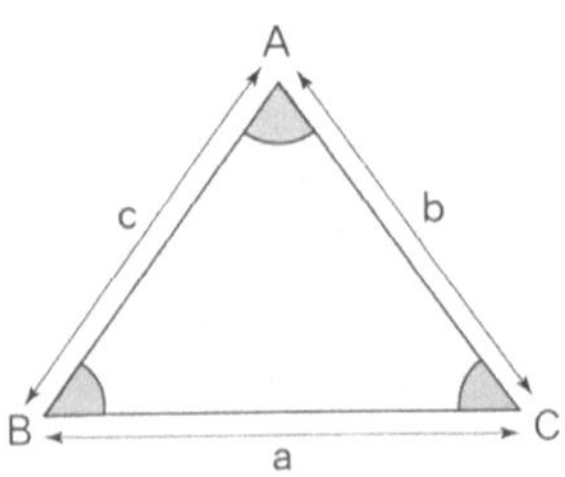

$$\frac{a}{\sin A} = \frac{b}{\sin B} = \frac{c}{\sin C}$$

Example: Given a = 20 units c = 25 units and Angle C = 42°. Find the angle A of the triangle.

Solution:

For the given data, we can use the following formula of sine law: a/sinA = b/sinB = c/sinC

$\Rightarrow$ 20/sin A = 25/sin 42°

$\Rightarrow$ sin A/20 = sin 42°/25

$\Rightarrow$ sin A = (sin 42°/25) × 20

$\Rightarrow$ sin A = (sin 42°/25) × 20

$\Rightarrow$ sin A = (0.6691/5) × 4

$\Rightarrow$ sin A = 0.5353

$\Rightarrow$ A = $\sin^{-1}$(0.5363)

$\Rightarrow$ A = 32.36°

Answer: "A = 32.36°

Proof of Law of Sines Formula

The law of sines is used to compute the remaining sides of a triangle, given two angles and a side. This technique is known as triangulation. It can also be applied when we are given two sides and one of the non-enclosed angles. But, in some such cases, the triangle cannot be uniquely determined by this given data, called the ambiguous case, and we obtain two possible values for the enclosed angle. To prove the sine law, we consider two oblique triangles as shown below.

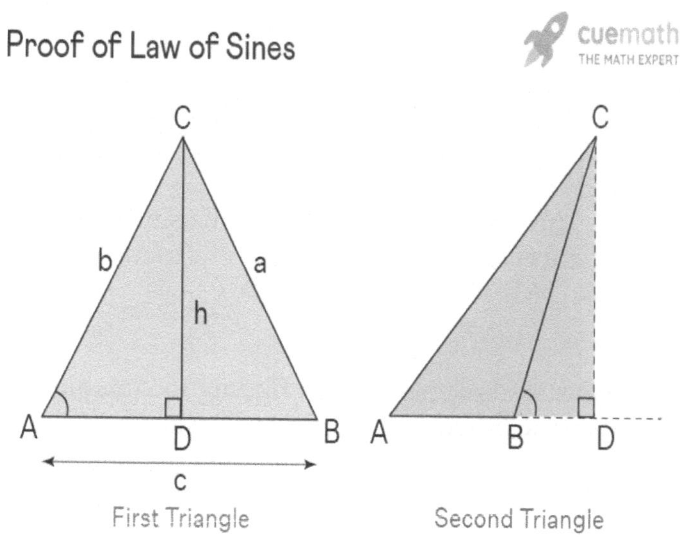

In the first triangle, we have:

h/b = sinA

⇒ h = b sinA

In the second triangle, we have:

h/a = sinB

⇒ h = a sinB

Also, sin(180° - B) = sinB

Equalizing the h values from the above expressions, we have:

a sinB = b sinA

⇒ a/sinA = b/sinB

Similarly, we can derive a relation for sin A and sin C.

asinC = csinA

⇒ a/sinA = c/sinC

Combining the above two expressions, we have the following sine law.

a/sinA = b/sinB = c/sinC

Tips and Tricks on Law of Sines

- The triangulation technique is used to find the sides of a triangle when two angles and one side of a triangle is known. For this the sine law is helpful.
- This sine law of trigonometry should not be confused with the sine law in physics.
- Further deriving from this sine law we can also find the area of an oblique triangle.
  Area of a triangle = (1/2) ab sinC = (1/2) bc sinA = (1/2) ca sinB
- Also sine law provides a relationship with the radius R of the circumcircle,a/sinA = b/sinB = c/sinC = 2R
- Cosine law: This proves a relationship between the sides and one angle of a triangle,$c^2 = a^2 + b^2 - 2ab.\cos C$
- Tangent law: This has been derived from the sine law and it gives the relationship between the sides and angles of a triangle. a–ba+b=tan(A–B)2tan(A–B)2a″ba+b=tan(A–B)2tan(A+B)2

Applications of Sine Law

The law of sines finds application in finding the missing side or angle of a triangle, given the other requisite data. The sine law can be applied to calculate:

- The length of the side of a triangle using ASA or AAS criteria.
- The unknown angle of a triangle.
- The area of the triangle.

Ambiguous Case of Law of Sines

While applying the law of sines to solve a triangle, there might be a case when there are two possible solutions, which occurs when two different triangles could be created using the given information. Let us understand this ambiguous case while solving a triangle using Sine law using the following example.

Example: If the side lengths ofΔABC are a = 18 and b = 20 with $\angle A$ opposite to 'a' measuring 26°, calculate the measure of $\angle B$ opposite to 'b'?

Solution:

Using the sine rule, we have $\sin A/a = \sin B/b = \sin 26^\circ/18 = \sin B/20$

$\Rightarrow \sin B = (9/10)\ \sin 26^\circ$ or $B \approx 29.149^\circ$.

However, note that $\sin x = \sin(180^\circ - x)$..$\therefore$ $A + B < 180^\circ$ and $A + (180^\circ - B) < 180^\circ$, another possible measure of B is approximately $180^\circ - 29.149^\circ = 150.851^\circ$.

Think out of the box:

Find the angles of a triangle if the sides are 12 units, 8.5 units, and 7.2 units respectively.

6

Cosine

Cosine is one of the primary mathematical trigonometric ratios. The ratio of the lengths of the side adjacent to the angle and the hypotenuse of a right-angled triangle is called the cosine function which varies as the angle varies. It is defined in the context of a right-angled triangle for acute angles. The cosine is used to model many real-life scenarios – radio waves, tides, sound waves, musical tones, electrical currents.

The cosine function is simply denoted as cos x, where x is the angle. In this article, we will learn the basic properties of the cosine, its graph, domain and range, derivative, integral, and its power series expansion of cosine. Cos x is a periodic function and has a period of 2Π.

WHAT IS COSINE?

Cosine or cos x is a periodic function in trigonometry. Consider a unit circle centered at the origin of the coordinate plane. A variable point P moves on the circumference of this circle. From the figure, we observe that P is in the first quadrant, and OP makes an acute angle of x radians with the positive x-axis. PQ is the perpendicular dropped from P onto the horizontal axis. The triangle is thus formed by joining the points O, P, and Q as shown in the figure, where OQ is the base, and PQ is the height of the triangle.

Hence, the cosine function for the above case can be mathematically written as:

cos x = OQ/OP, Here, x is the acute angle formed between the hypotenuse and the base of a right-angled triangle.

Cosine Meaning

The cosine can be defined as the ratio of the length of the base to the length of the hypotenuse in a right-angled triangle. Mathematically, the cosine function formula in terms of sides of a right-angled triangle is written as:

cos x = Adjacent Side/Hypotenuse = Base/Hypotenuse, where x is the acute angle between the base and the hypotenuse.

Cosine Function from Unit Circle

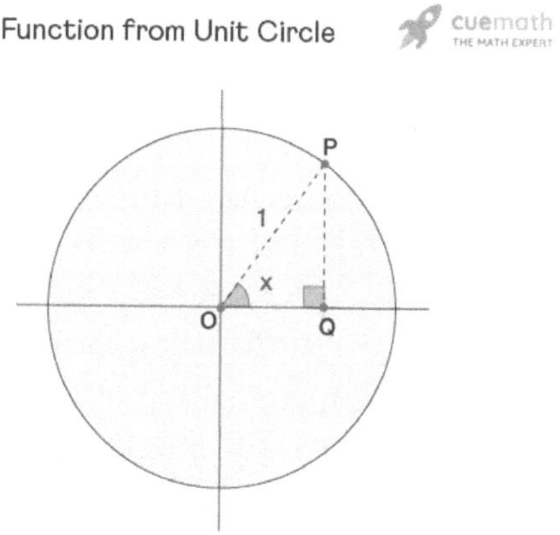

Cosine Graph

As shown in the image above, we note that cos x = OQ/OP = OQ/1 = OQ. As x varies, the value of the cosine varies with the variation in the length of OQ. Now, we will study the variation in the cosine function in the four-quadrant of the coordinate plane.

Case 1: Variation of OQ in the first quadrant.

Suppose that initially, P is on the horizontal axis. Let us consider a movement of P through 90° or Π/2 rad. The following figure shows different positions of Q for this movement. Clearly, OQ has decreased in length, from an initial value of 1 (when x is 0 radians) to a final value of 0 (when x is Π/2 radians).

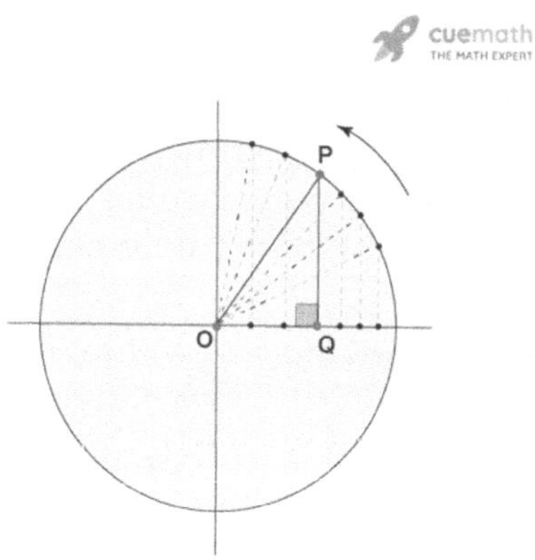

Case 2: Variation of OQ in the second quadrant.

Now, we will check the position of P in the second quadrant as we did in the first quadrant and check how the value of the cosine function varies. P subsequently moves from 90° position to 180° position. In this phase of the movement, the length or magnitude of OQ increases, and the value of cosine has decreased from a value of 0 at 90°, to a minimum of -1 at 180°.

Case 3: Variation of OQ in the third quadrant.

When P moves from a position of 180° to a position of 270°, though the length or magnitude of OQ decreases. But since the direction is along the negative y-axis, the actual value of cos x increases from -1 to 0. Thus, the value of the cosine for angle x increases.

Case 4: Variation of OQ in the fourth quadrant.

Finally, when P moves from a position of 270° to a position of 360°, OQ increases from 0 to 1 (once again). The length or magnitude of OQ increases along with the algebraic value of OQ increasing. Thus, the value of the cosine function for angle x increases.

We can now plot this variation on a graph. The horizontal axis represents the input variable x as the angle in radians, and the vertical axis represents the value of the cosine function for x. Merging the response of variation in the value of PQ for all four quadrants, we obtained the complete plot of cos x vs x, for one complete cycle of 0 radians to 2Π radians (0° to 360°). The plot thus obtained is shown below:

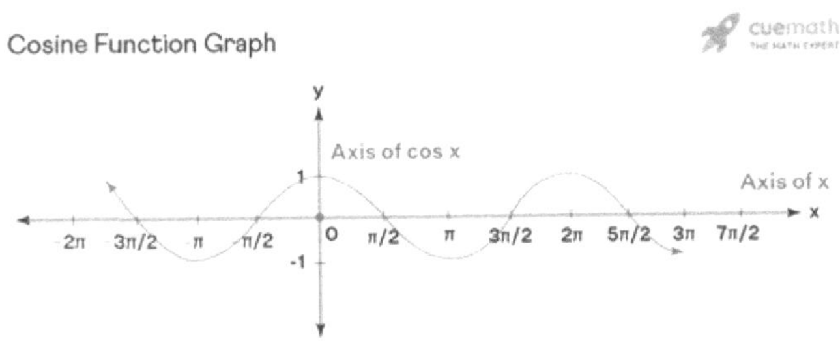

Cosine Values

We study the value of the cosine function for some specific angles as they are easy to remember. These cosine values are used in solving different mathematical problems. Some of these values of the cosine are listed below in the trigonometric table:

Cosine Degrees	Cosine Radians	Value of Cosine Function (cos x)
cos 0°	cos 0	1
cos 30°	cos Π/6	√3/2
cos 45°	cos Π/4	1/√2

cos 60°	cos Π/3	1/2
cos 90°	cos Π/2	0
cos 120°	cos 2Π/3	-1/2
cos 150°	cos 5Π/6	-√3/2
cos 180°	cos Π-1	
cos 270°	cos 3Π/2	0
cos 360°	cos 2Π	1

Properties of Cosine Function

Properties of cosine depend upon the quadrant in which the angle lies. The cosine function is a special trigonometric function and has many properties. Some of them are listed below:

- The cos x graph repeats itself after 2π, which suggests the function is periodic with a period of 2π.
- Cos x is an even function because cos(−x) = cos x.
- The domain of cosine function is all real numbers and the range is [-1,1].
- The reciprocal of the cosine function is the secant function.
- Power series expansion of the cosine is cos x = Σ∞n=0(−1)nx2n(2n)!Σn=0∞(−1)nx2n(2n)!

Cosine Function Identities

In trigonometry, there are several identities involving the cosine function. These identities are very useful in solving various math problems. Some of them are listed below:

- cos x = 1/ sec x
- Inverse of cosine function = $\cos^{-1}x$ = arccos x, where x lies in.
- $\sin^2x + \cos^2x = 1$
- cos (x + y) = cos x cos y - sin x sin y
- cos (x - y) = cos x cos y + sin x sin y
- $\cos 2x = \cos^2x - \sin^2x = 2\cos^2x - 1 = 1 - 2\sin^2x$
- Derivative of cos x: d(cos x)/dx = -sin x
- Integral of cosine function: ∫cos x dx = sin x + C, where C is the constant of integration.

7

Similar Triangles

Similar triangles are the triangles that have corresponding sides in proportion to each other and corresponding angles equal to each other. Similar triangles look the same but the sizes can be different. In general, similar triangles are different from congruent triangles. There are various methods by which we can find if two triangles are similar or not. Let us learn more about similar triangles and their properties along with a few solved examples.

WHAT ARE SIMILAR TRIANGLES?

Similar triangles are the triangles that look similar to each other but their sizes might not be exactly the same. Two objects can be said similar if they have the same shape but might vary in size. That means similar shapes when magnified or demagnified superimpose each other. This property of similar shapes is referred to as "Similarity".

Similar Triangles Definition

Two triangles will be similar if the angles are equal (corresponding angles) and sides are in the same ratio or proportion(corresponding sides). Similar triangles may have different individual lengths of the sides of triangles but their angles must be equal and their corresponding ratio of the length of the sides must be the same. If two triangles are similar that means,

- All corresponding angle pairs of triangles are equal.
- All corresponding sides of triangles are proportional.

We use the "~" symbol to represent the similarity. So, if two triangles are similar, we show it as$\Delta QPR \sim \Delta XYZ$

SIMILAR TRIANGLES EXAMPLES

Similar triangles are triangles for which the corresponding angle pairs are equal. That means equiangular triangles are similar. Therefore, all equilateral triangles are examples of similar triangles. The following image shows similar triangles, but we must notice that their sizes are different.

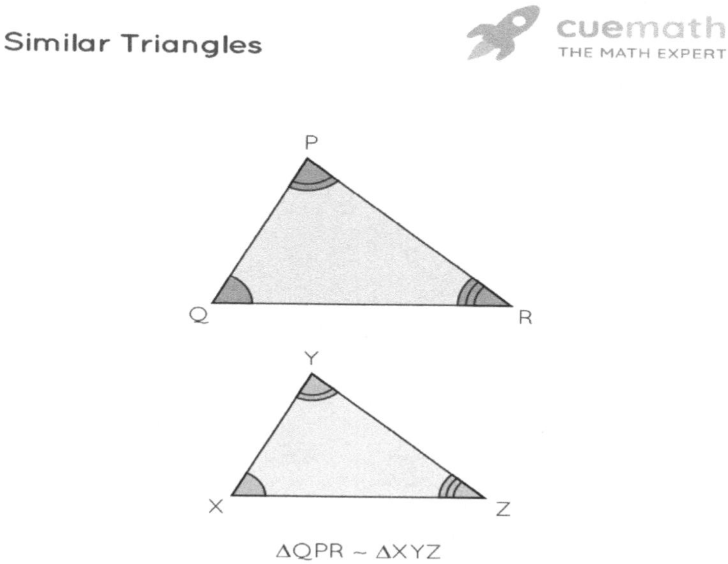

Similar Triangles Formulas

In the previous section, we saw there are two conditions using which we can verify if the given set of triangles are similar or not. These conditions state that two triangles can be said similar if either their corresponding angles are equal or congruent or if their corresponding sides are in proportion. Therefore, two triangles ΔABC and ΔEFG can be proved similar(ΔABC ~ΔEFG) using either condition among the following set of similar triangles formulas,

Formula for Similar Triangles in Geometry:

- $\angle A = \angle E$, $\angle B = \angle F$ and $\angle C = \angle G$
- $AB/EF = BC/FG = AC/EG$

Similar Triangles Theorems

We can find out or prove whether two triangles are similar or not using the similarity theorems. We use these similarity criteria when we do not have the measure of all the sides of the triangle or measure of all the angles of the triangle. These similar triangle theorems help us quickly find out whether two triangles are similar or not. There are three major types of similarity rules, as given below,

- AA (or AAA) or Angle-Angle Similarity Theorem
- SAS or Side-Angle-Side Similarity Theorem
- SSS or Side-Side-Side Similarity Theorem

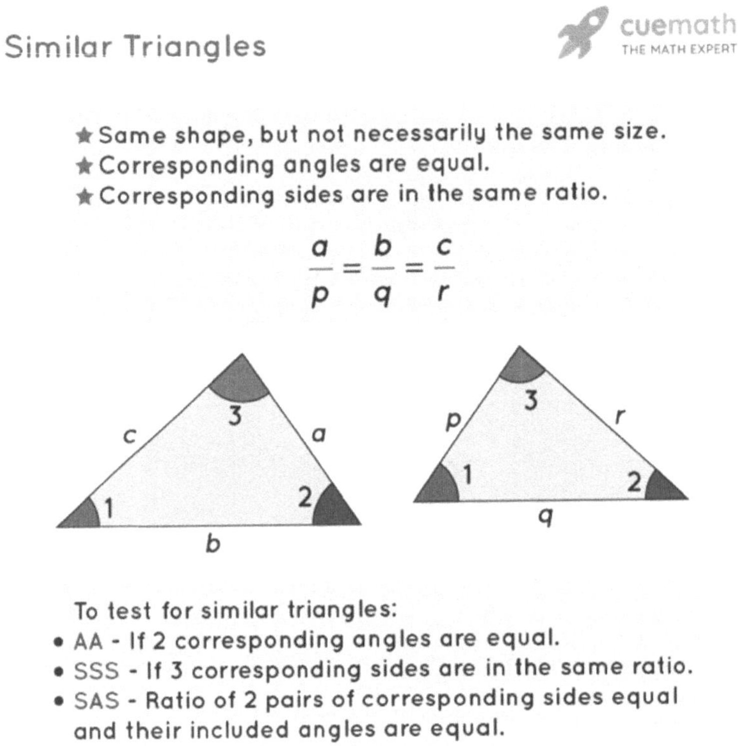

Let us understand these similar triangles theorems with their proofs.

AA (or AAA) or Angle-Angle Similarity Criterion

AA similarity criterion states that if any two angles in a triangle are respectively equal to any two angles of another triangle, then they must be similar triangles. AA similarity rule is easily applied when we only know the measure of the angles and have no idea about the length of the sides of the triangle. In the image given below, if it is known that $\angle B = \angle G$, and $\angle C = \angle F$.

And we can say that by the AA similarity criterion,ΔABC andΔEGF are similar or$\Delta ABC \sim \Delta EGF$.

$\Rightarrow AB/EG = BC/GF = AC/EF$ and $\angle A = \angle E$.

Click here to understand AA Similarity Criterion in detail- AA similarity criterion

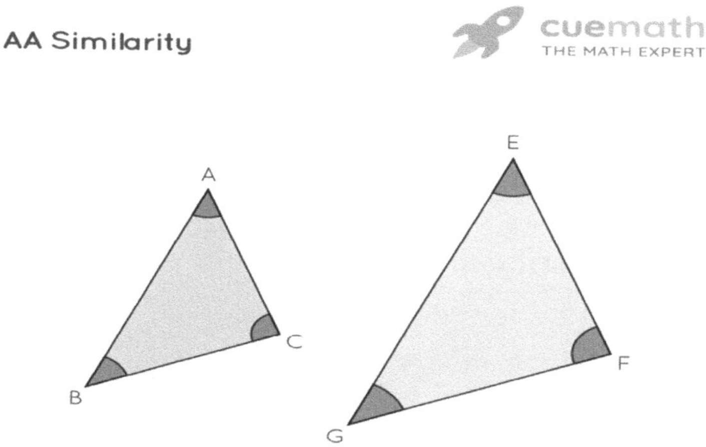

SAS or Side-Angle-Side Similarity Criterion

According to the SAS similarity theorem, if any two sides of the first triangle are in exact proportion to the two sides of the second triangle along with the angle formed by these two sides of the individual triangles are equal, then they must be similar triangles. This rule is generally applied when we only know the measure of two sides and the angle formed between those two sides in both the triangles respectively.

In the image given below, if it is known that AB/DE = AC/DF, and ∠A = ∠D

And we can say that by the SAS similarity criterion,ΔABC andΔDEF are similar orΔABC~ΔDEF.

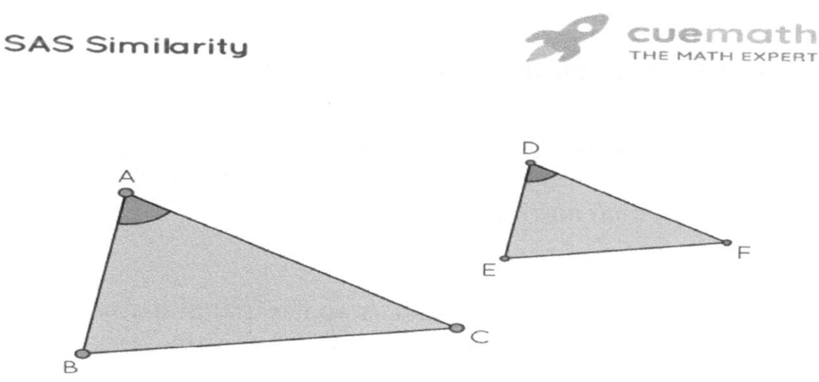

SSS or Side-Side-Side Similarity Criterion

According to the SSS similarity theorem, two triangles will the similar to each other if the corresponding ratio of all the sides of the two triangles are equal. This criterion is commonly used when we only have the measure of the sides of the triangle and have less information about the angles of the triangle.

In the image given below, if it is known that PQ/ED = PR/EF = QR/DF

And we can say that by the SSS similarity criterion,ΔPQR andΔEDF are similar orΔPQR~ΔEDF.

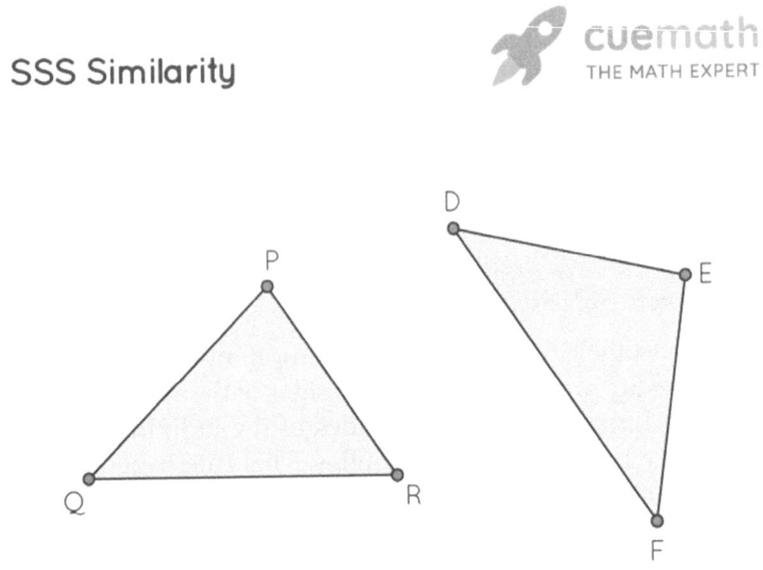

Similar Triangles Properties

If two triangles are similar or proved similar by any of the above-stated criteria, then they possess few properties of the similar triangles. Properties of similar triangles are given below,

- Similar triangles have the same shape but different sizes.
- In similar triangles, corresponding angles are equal.
- Corresponding sides of similar triangles are in the same ratio.
- The ratio of area of similar triangles is the same as the ratio of the square of any pair of their corresponding sides.

How to Find Similar Triangles?

Two given triangles can be proved as similar triangles using the above-given theorems. We can follow the steps given below to check if the given triangles are similar or not,

- **Step 1:** Note down the given dimensions of the triangles (corresponding sides or corresponding angles).
- **Step 2:** Check if these dimensions follow any of the conditions for similar triangles theorems(AA, SSS, SAS).

- **Step 3:** The given triangles, if satisfy any of the similarity theorems, can be represented using the "~" to denote similarity.

Let us understand these steps better using an example.ifΔABC andΔPQR are similar triangles or not using the given data: $\angle$A = 65°, $\angle$B = 70° and $\angle$P = 70°, "R = 45°.

Solution:

Using the given measurement of angles, we cannot conclude if the given triangles follow the AA similarity criterion or not. Let us find the measure of the third angle and evaluate.

We know, using angle sum property of a triangle, $\angle$C inΔABC = 180° - ($\angle$A + $\angle$B) = 180° - 135° = 45°

Similarly, $\angle$Q inΔPQR = 180° - ($\angle$P + $\angle$R) = 180° - 115° = 65°

Therefore, we can conclude that inΔABC andΔPQR, $\angle$A = $\angle$Q, $\angle$B = $\angle$P, and $\angle$C = R

$\Rightarrow\Delta ABC \sim \Delta QPR$

DIFFERENCE BETWEEN SIMILAR TRIANGLES AND CONGRUENT TRIANGLES

Similarity and congruency are two different properties of triangles. The following table helps in distinguishing similar triangles with congruent triangles: Similar Triangles Congruent Triangles

Similar triangles have the same shape but may be different in size. They superimpose each other when magnified or demagnified. Congruent triangles are the same in shape and size. They superimpose each other in their original shape.

They are represented using the symbol is '~'. For example, Similar triangles ABC and XYZ will be represented as,ΔABC $\cong\Delta$XYZ They are represented using the symbol is 'E"'. For example, Congruent triangles ABC and XYZ will be represented as,ΔABC $\cong$ ΔQPR

The ratio of all the corresponding sides is equal in similar triangles. This common ratio is also called as 'scale factor' in similar triangles.
The ratio of corresponding sides is equal to 1 for congruent triangles.

8

Pythagoras Theorem

The Pythagoras theorem which is also referred to as the Pythagorean theorem explains the relationship between the three sides of a right-angled triangle. According to the Pythagoras theorem, the square of the hypotenuse is equal to the sum of the squares of the other two sides of a triangle. Let us learn more about the Pythagoras theorem, its derivations, and equations followed by solved examples on the Pythagoras theorem triangle and squares.

WHAT IS THE PYTHAGORAS THEOREM?

The Pythagoras theorem states that if a triangle is right-angled (90 degrees), then the square of the hypotenuse is equal to the sum of the squares of the other two sides. Observe the following triangle ABC, in which we have $BC^2 = AB^2 + AC^2$. Here, AB is the base, AC is the altitude (height), and BC is the hypotenuse. It is to be noted that the hypotenuse is the longest side of a right-angled triangle.

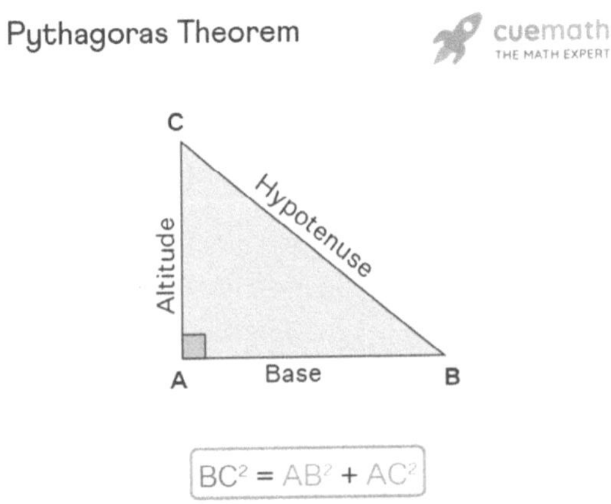

Pythagoras Theorem Equation

The Pythagoras theorem equation is expressed as, $c^2 = a^2 + b^2$, where 'c' = hypotenuse of the right triangle and 'a' and 'b' are the other two legs. Hence, any triangle with one angle equal to 90 degrees produces a Pythagoras triangle and the Pythagoras equation can be applied in the triangle.

History of Pythagoras Theorem

Pythagoras theorem was introduced by the Greek Mathematician Pythagoras of Samos. He was an ancient Ionian Greek philosopher. He formed a group of mathematicians who works religiously on numbers and lived like monks. Finally, the Greek Mathematician stated the theorem hence it was named after him as the "Pythagoras theorem." Though it was introduced many centuries ago its application in the current era is obligatory to deal with pragmatic situations.

Although Pythagoras introduced and popularised the theorem, there is sufficient evidence proving its existence in other civilizations, 1000 years before Pythagoras was born. The oldest known evidence dates back to between 20th to 16th Century B.C in the Old Babylonian Period.

Pythagorean Theorem Formula

The Pythagoras theorem formula states that in a right triangle ABC, the square of the hypotenuse is equal to the sum of the square of the other two legs. If AB and AC are the sides and BC is the hypotenuse of the triangle, then: $BC^2 = AB^2 + AC^2$. In this case, AB is the base, AC is the altitude or the height, and BC is the hypotenuse.

Another way to understand the Pythagorean theorem formula is using the following figure which shows that the area of the square formed by the longest side of the right triangle (the hypotenuse) is equal to the sum of the area of the squares formed by the other two sides of the right triangle.

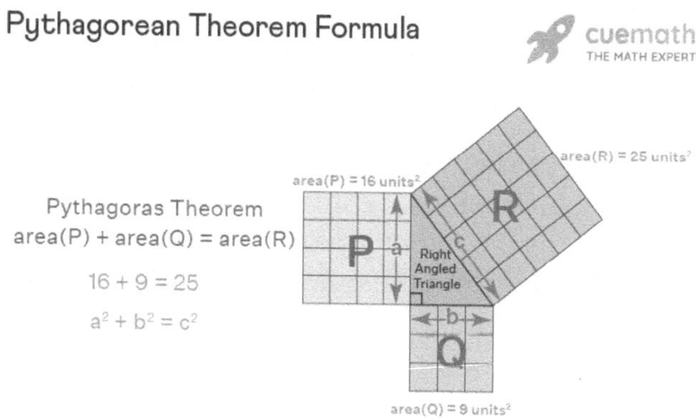

In a right-angled triangle, the Pythagoras Theorem Formula is expressed as:

$c^2 = a^2 + b^2$

Where,

- 'c' = hypotenuse of the right triangle
- 'a' and 'b' are the other two legs.

Pythagoras Theorem Proof

Pythagoras theorem can be proved in many ways. Some of the most common and widely used methods are the algebraic method and the similar triangles method. Let us have a look at both these methods individually in order to understand the proof of this theorem.

Proof of Pythagorean Theorem Formula using the Algebraic Method

The proof of the Pythagoras theorem can be derived using the algebraic method. For example, let us use the values a, b, and c as shown in the following figure and follow the steps given below:

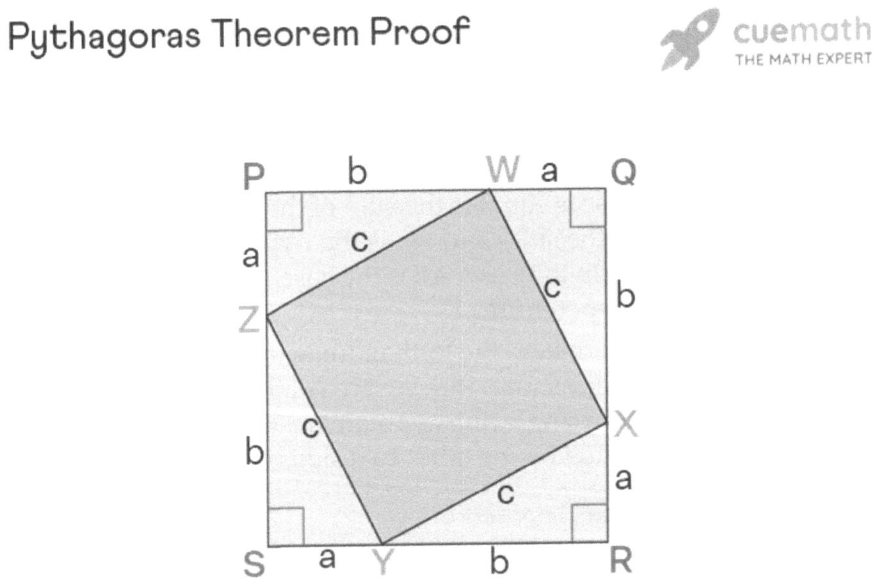

- **Step 1:** Arrange four congruent right triangles in the given square PQRS, whose side is a + b. The four right triangles have 'b' as the base, 'a' as the height and, 'c' as the hypotenuse.
- **Step 2:** The 4 triangles form the inner square WXYZ as shown, with 'c' as the four sides.
- **Step 3:** The area of the square WXYZ by arranging the four triangles is c^2.

- **Step 4:** The area of the square PQRS with side (a + b) = Area of 4 triangles + Area of the square WXYZ with side 'c'. This means $(a + b)^2 = [4 \times 1/2 \times (a \times b)] + c^2$.This leads to $a^2 + b^2 + 2ab = 2ab + c^2$. Therefore, $a^2 + b^2 = c^2$. Hence proved.

Pythagorean Theorem Formula Proof using Similar Triangles

Two triangles are said to be similar if their corresponding angles are of equal measure and their corresponding sides are in the same ratio. Also, if the angles are of the same measure, then by using the sine law, we can say that the corresponding sides will also be in the same ratio. Hence, corresponding angles in similar triangles lead us to equal ratios of side lengths.

Derivation of Pythagorean Theorem Formula

Consider a right-angled triangle ABC, right-angled at B. Draw a perpendicular BD meeting AC at D.

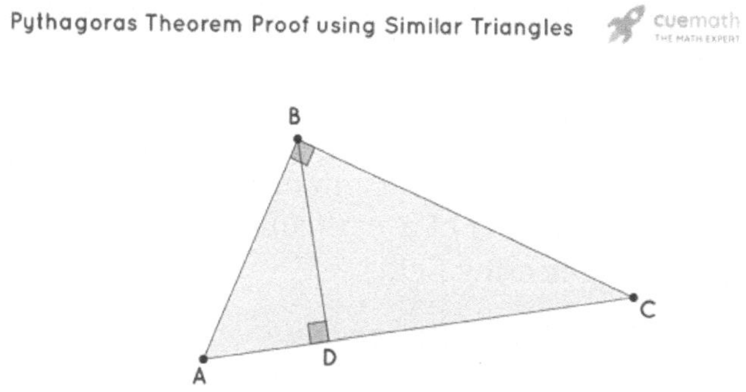

InΔABD andΔACB,

- $\angle A = \angle A$ (common)
- $\angle ADB = \angle ABC$ (both are right angles)

Thus,$\Delta ABD \sim \Delta ACB$ (by AA similarity criterion)

Similarly, we can prove$\Delta BCD \sim \Delta ACB$.

Thus$\Delta ABD \sim \Delta ACB$, Therefore, AD/AB = AB/AC. We can say that $AD \times AC = AB^2$.

Similarly,$\Delta BCD \sim \Delta ACB$. Therefore,CD/BC = BC/AC. We can also say that $CD \times AC = BC^2$.

Adding these 2 equations, we get $AB^2 + BC^2 = (AD \times AC) + (CD \times AC)$

$AB^2 + BC^2 = AC(AD + DC)$

$AB^2 + BC^2 = AC^2$

Hence proved.

Pythagoras Theorem Triangles

Right triangles follow the rule of the Pythagoras theorem and they are called Pythagoras theorem triangles. The three sides of such a triangle are collectively called Pythagoras triples. All the Pythagoras theorem triangles follow the Pythagoras theorem which says that the square of the hypotenuse is equal to the sum of the two sides of the right-angled triangle. This can be expressed as $c^2 = a^2 + b^2$; where 'c' is the hypotenuse and 'a' and 'b' are the two legs of the triangle.

Pythagoras Theorem Squares

As per the Pythagorean theorem, the area of the square which is built upon the hypotenuse of a right triangle is equal to the sum of the area of the squares built upon the other two sides. These squares are known as Pythagoras squares.

Pythagoras Theorem

ns of the Pythagoras theorem can be seen in our day-to-day life. Here are some of the applications of the Pythagoras theorem.

- Engineering and Construction fields

Most architects use the technique of the Pythagorean theorem to find the unknown dimensions. When length or breadth are known it is very easy to calculate the diameter of a particular sector. It is mainly used in two dimensions in engineering fields.

- *Face recognition in security cameras*

The face recognition feature in security cameras uses the concept of the Pythagorean theorem, that is, the distance between the security camera and the location of the person is noted and well-projected through the lens using the concept.

- *Woodwork and interior designing*

The Pythagoras concept is applied in interior designing and the architecture of houses and buildings.

- *Navigation*

People traveling in the sea use this technique to find the shortest distance and route to proceed to their concerned places.

9

Hypotenuse Leg Theorem

In a right-angled triangle, the hypotenuse is the longest side which is always opposite to the right angle. The hypotenuse leg theorem states that two right triangles are congruent if the hypotenuse and one leg of one right triangle are congruent to the other right triangle's hypotenuse and leg side. In order to prove any two right triangles congruent, we apply the HL (Hypotenuse Leg) Theorem or the RHS (Right angle-Hypotenuse-Side) congruence rule. Let us learn more about the hypotenuse leg theorem in this page.

HYPOTENUSE LEG THEOREM

According to the hypotenuse leg theorem, if the hypotenuse and one leg of one right triangle are congruent to the other right triangle's hypotenuse and leg side, then the two triangles are congruent.

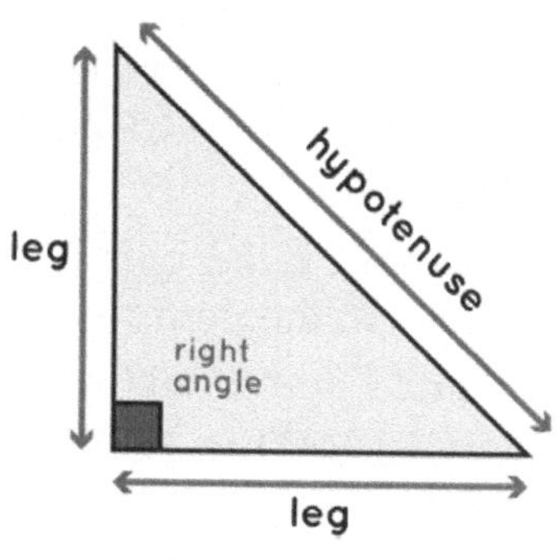

In other words, a given set of right triangles are congruent if the corresponding lengths of their hypotenuse and one leg are equal. The hypotenuse leg theorem is a criterion that is used to prove the congruence of triangles. In the other congruency postulates like, Side Side Side (SSS), Side Angle Side (SAS), Angle Side Angle (ASA), and Angle Angle Side (AAS), three criteria are tested, whereas, in the hypotenuse leg (HL) theorem, only the hypotenuse and one leg are considered. Observe the following figure which shows a right-angled triangle with two perpendicular legs and a hypotenuse.

Hypotenuse Leg Theorem Proof

The proof of the hypotenuse leg theorem shows how a given set of right triangles are congruent if the corresponding lengths of their hypotenuse and one leg are equal. Observe the following isosceles triangle ABC in which side AB = AC and AD is perpendicular to BC.

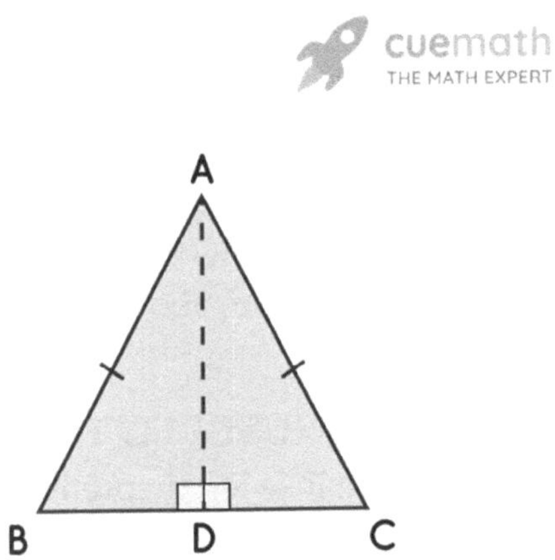

Given: Here, ABC is an isosceles triangle, AB = AC, and AD is perpendicular to BC.

Proof:

AD, being an altitude is perpendicular to BC and forms ADB and ADC as right-angled triangles. AB and AC are the respective hypotenuses of these triangles, and we know they are equal to each other. AD = AD because they are common in both the triangles.

So, AB = AC and AD is common.

Therefore, a hypotenuse and a leg pair in two right triangles, are satisfying the definition of the HL theorem.

We know that angles B and C are equal (Isosceles Triangle Property).

We also know that the angles BAD and CAD are equal.(AD bisects BC, which makes BD equal to CD).

Therefore,$\Delta ADB \sim \Delta ADC$

Hence proved.

Important Notes

- The Pythagorean theorem states that in a right-angled triangle, the square of the hypotenuse (longest side) is equal to the sum of the squares of the other two sides (base and perpendicular). This is represented as: $\text{Hypotenuse}^2 = \text{Base}^2 + \text{Perpendicular}^2$.
- According to the HL Congruence rule, the hypotenuse and one leg are the elements that are used to test the congruence of triangles.
- The HL Congruence rule is similar to the SAS (Side-Angle-Side) postulate. The only difference is that SAS needs two sides and the included angle, whereas, in the HL theorem, the known angle is the right angle, which is not the included angle between the hypotenuse and the leg.

RIGHT ANGLED TRIANGLE CONSTRUCTIONS(RHS)

A triangle is a closed shape that has 3 sides. A right-angled triangle is a type of triangle having the 3 sides, named as the 'base', 'height', and the 'hypotenuse'. Right-angled triangles are extensively used in trigonometry. A right-angled triangle has one angle equal to 90° and the other two angles as acute. The side that is opposite to the right angle is called the 'hypotenuse', which is also the longest side of the right-angled triangle. The side that is adjacent to the right angle is called the 'adjacent side'. The height of the triangle forms the 'opposite side'.

Right Angled Triangle Definition

A triangle that has one of its internal angles equal to 90° is called a right-angled triangle. The longest side of a right-angled triangle is called the hypotenuse. The length of the longest side of the right-angled triangle can be found using the pythagoras theorem, which states that 'The sum of the square of the hypotenuse is equal to the sum of the squares of the other two sides.'

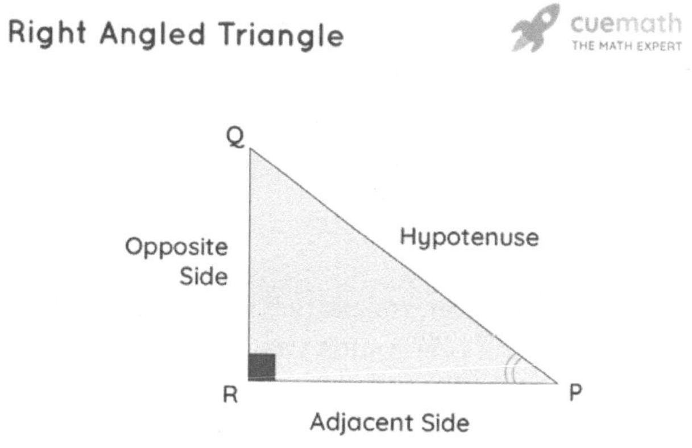

Right Angled Triangle Properties

The following points are the properties of a right-angled triangle.

- A right-angle triangle has one of its angles equal to 90°
- The longest side of a right-angle triangle is called 'Hypotenuse'.
- The sum of interior angles of a right-angled triangle is equal to 180°
- The side that is opposite to the right angle is called the opposite side and the side that is next to the right angle is the adjacent side.
- The area of a right-angled triangle is equal to (1/2 × base × height)
- If any two sides of a right-angled triangle are given we can construct the triangle easily.
- One of the unknown sides of a right-angled triangle can be found using the pythagorean theorem formula - Hypotenuse = $\sqrt{\text{Opposite side}^2 + \text{Adjacent side}^2}$

Construction of a Right Angled Triangle

A triangle in which one of the angles is equal to 90° is a right-angled triangle. The side which is directly opposite to the right angle is called the hypotenuse of the longest side. To construct a right-angled triangle, we require the measurements of two of its sides. A compass and a ruler are necessary to construct a right-angled triangle. Now let us see how a triangle PQR with the hypotenuse13 units and one of its sides to be 5 units.

Step 1: Draw a horizontal line and mark a point Q on it. The line can be of any length.

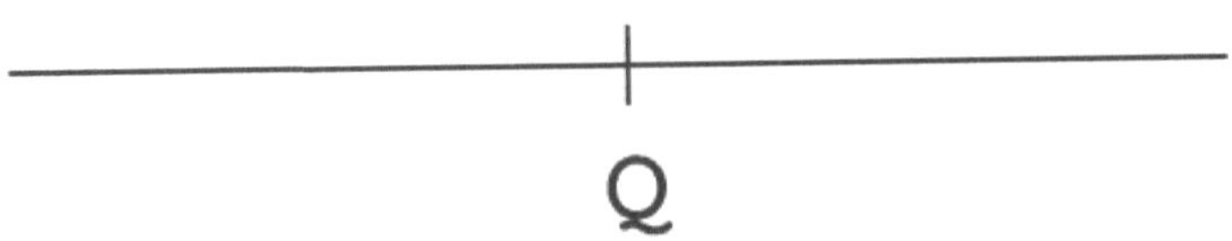

Step 2: With Q as the center, measure 5 units in a compass and draw an arc on both the sides of the point such that the arc touches the horizontal line and mark the points as 'S' and 'R'.

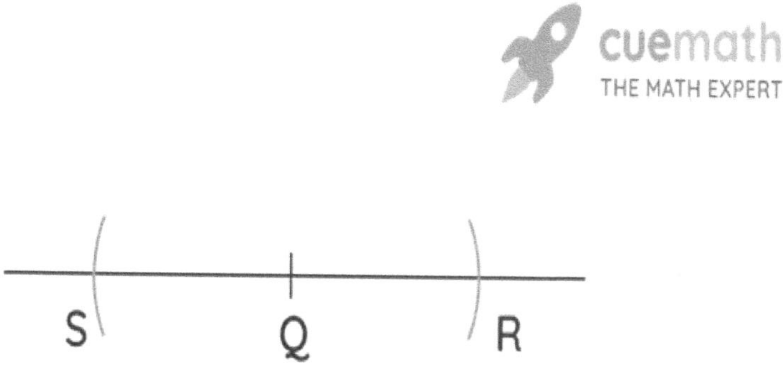

Step 3: With 'S' as the center and measuring 13 units in the compass draw an arc above 'S'.

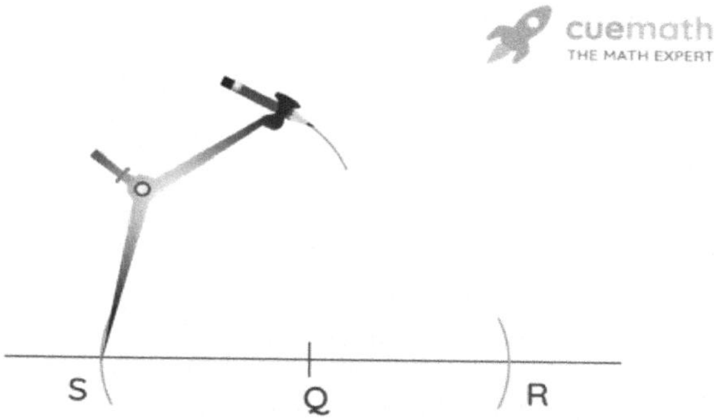

Step 4: With the same width of 13 units, draw an arc from the point 'R'. Mark the point of intersection of these arcs as 'P'.

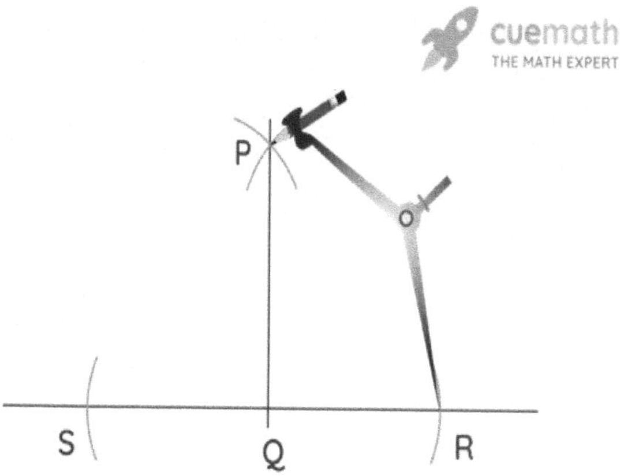

Step 5: Join the points 'P' and 'Q' and 'P' and 'R' with a ruler.

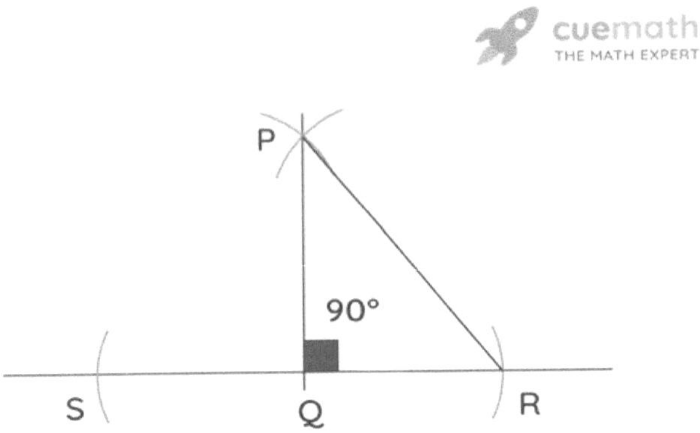

Step 6: The angle at point Q is 90°.

Right Angled Triangle

A right angled triangle is a triangle with one of the angles as 90 degrees. A 90-degree angle is called a right angle, and hence the triangle with a right angle is called a right triangle. In this triangle, the relationship between the various sides can be easily understood with the help of the Pythagoras rule. The side opposite to the right angle is the largest side and is referred to as the hypotenuse. Further, based on the other angle values, the right triangles are classified as an isosceles right triangle and a scalene right triangle. Also, the lengths of the sides of the right triangle, such as 3, 4, 5 are referred to as Pythagorean triples.

What is a Right Triangle?

The definition for a right triangle states that if one of the angles of a triangle is a right angle - 90°, the triangle is called a right-angled triangle or simply, a right triangle. In the given image, triangle ABC is a right triangle, where we have the base, the altitude, and the hypotenuse. Here AB is the base, AC is the altitude, and BC is the hypotenuse. The hypotenuse is the important side of a right triangle which is the largest side and is opposite to the right angle within the triangle.

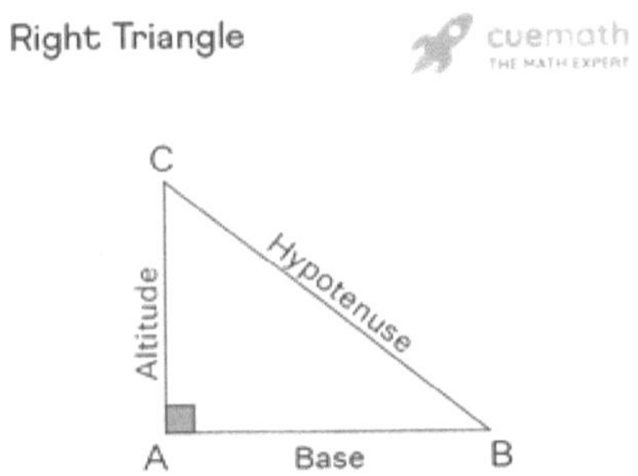

Here we can have understood the distinct features of a right triangle. The features of triangle ABC are as follows:

- AC is the height, altitude, or perpendicular
- AB is the base
- AC ⊥ AB
- ∠A=90°
- The side BC opposite to the right angle is called the hypotenuse and it is the longest side of the right triangle.

Some of the examples of right triangles in our daily life are the triangular slice of bread, a square piece of paper folder across the diagonal, or the 30-60-90 triangular scale in a geometry box.

Right Triangle Formula

The great Greek philosopher, Pythagoras, derived an important formula for a right triangle. The formula states that in a right triangle, the square of the hypotenuse is equal to the sum of the squares of the other two legs. It was named after him as Pythagoras theorem. The right triangle formula can be represented in the following way: The square of the hypotenuse is equal to the sum of the square of the base and the square of the altitude.

In a right triangle we have: $(\text{Hypotenuse})^2 = (\text{Base})^2 + (\text{Altitude})^2$

Pythagorean Triplet: The three numbers which satisfy the above equation are the Pythagorean triplets. For example, (3, 4, 5) is a Pythagorean triplet because we know that $3^2= 9$, $4^2 = 16$, and $5^2 = 25$ and, $9 +16 = 25$. Therefore, $3^2 + 4^2 = 5^2$ These three numbers satisfying this condition are called the Pythagorean triplet. Some of the other examples of Pythagorean triples are (6, 8, 10), and (12, 5, 13).

PERIMETER OF A RIGHT TRIANGLE

The right triangle perimeter is the sum of the measures of all the 3 sides. It is the sum of the base, altitude, and hypotenuse of the right triangle. Here, for the below right triangle, the perimeter is equal to the sum of the sides BC + AC + AB = (a + b + c) units. The perimeter is a linear value and has a unit of length.

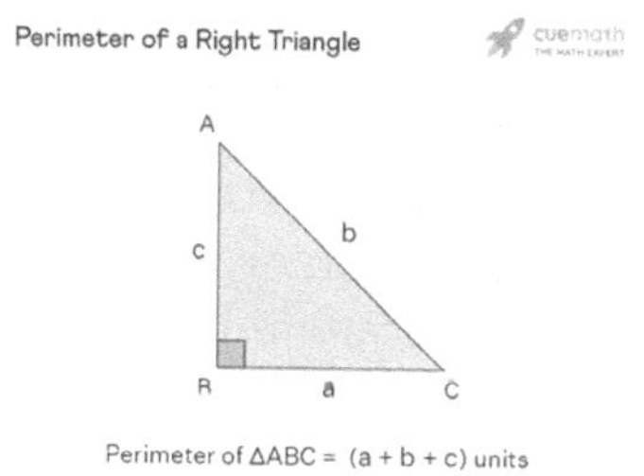

Right Triangle Area

The area of a right triangle gives the spread or the space occupied by the triangle. It is equal to half of the product of the base and the height of the triangle. It is a two-dimensional quantity and therefore represented in square units. The only two sides needed to find the right-angled triangle area are the base and the altitude.

Applying the right triangle definition, the area of a right triangle is given by the formula: Area of a right triangle = (1/2 × base × height) square units.

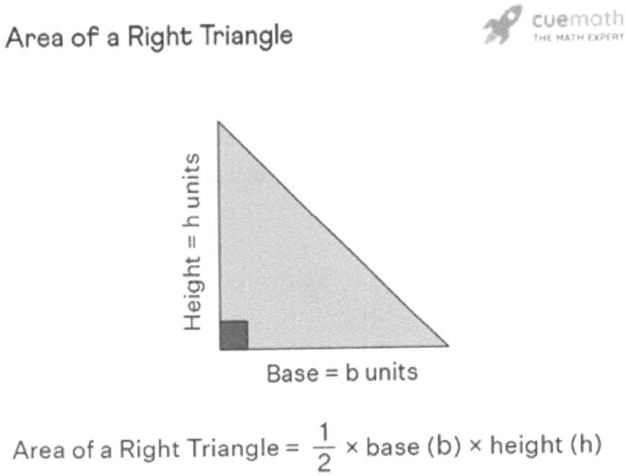

PROPERTIES OF RIGHT TRIANGLE

The first property of a right triangle is that it has one of its angles as 90°. The 90° angle is a right angle and the largest angle of a right triangle. Also, the other two angles are lesser than 90° or are acute angles. The right triangle properties are listed below:

- The largest angle is always 90°.
- The largest side is called the hypotenuse which is always the side opposite to the right angle.
- The measurements of the sides follow the Pythagoras rule.
- It cannot have any obtuse angle.

Types of Right Triangles

We have learned that one of the angles in a right triangle is 90°. This implies that the other two angles in the triangle will be acute angles. There are a few special right triangles namely the isosceles right triangles and the scalene right triangles. The triangle having both the other two angles equal is referred to as an isosceles right triangle, and the triangle with the other two angles having different values is called a scalene right triangle.

Isosceles Right Triangle

An isosceles right triangle is called a 90°-45°- 45° triangle. In triangle ABC, angle A = 90°; so by right triangle definition, triangle ABC is a right triangle. Also AB = AC; since two sides are equal, the triangle is also an isosceles triangle. Since AB = AC, the base angles are equal. We know that the sum of the angles of a triangle is 180°. Hence, the base angles add up to 90° which implies that they are 45° each. So in an isosceles right triangle, angles will always be 90°-45°- 45°.

Scalene Right Triangle

A scalene right triangle is a triangle where one angle is 90° and the other two angles that up to 90° are of different measurements. In the triangle PQR, ∠Q =90°, hence, it is a right triangle. PQ is not equal to QR, hence, it is a scalene triangle. There is also a special case of a scalene triangle 30°-60°-90° which is also a right triangle where the ratio of the triangle's longest side to its shortest side is 2:1. The side opposite to the 30° angle is the shortest side.

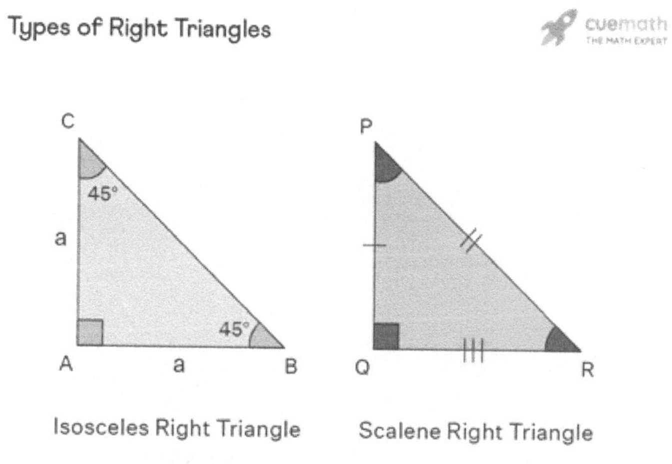

Tips & Tricks

Listed here are some of the important tips and tricks relating to a right triangle.

- The measurements of the side lengths will always satisfy the Pythagoras theorem.
- In a right triangle, the hypotenuse is the side opposite to the right angle and it is the longest side of the triangle.
- The other two legs are perpendicular to each other; one is the base and the other is the height.

Right Triangle Formulas

A triangle is a closed figure or shape with 3 sides, 3 angles, and 3 vertices, and for right triangle formulas, the properties have to be more specific. If any one of the angles of a triangle is a right angle (measuring 90°), the triangle is called a right-angled triangle or simply, a right triangle. Right triangle formulas would help you solve various calculations related to the perimeter, area, etc of the right triangle.

What Are the Right Triangle Formulas?

A right-angled triangle is one which has one of its interior angle measuring 90 degrees. Right-angled triangle formulas are used to calculate the perimeter, area, height, etc of a right triangle using its three sides.

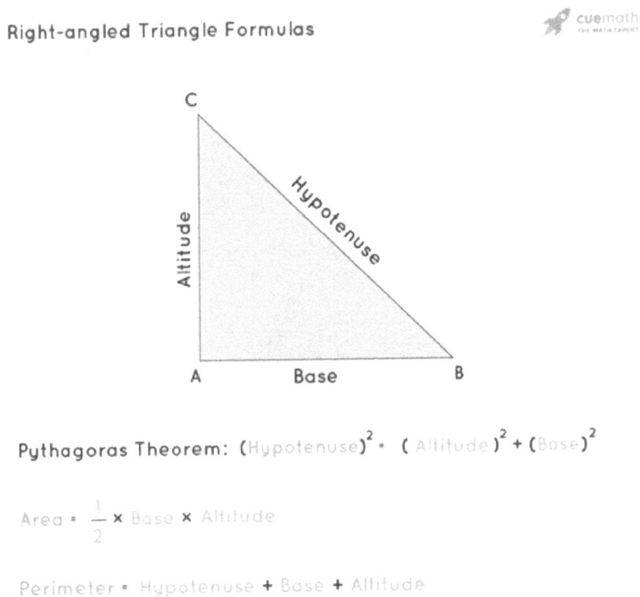

RIGHT-ANGLED TRIANGLE FORMULA

Different formulas associated with the right triangle are:

- Pythagoras Theorem - Formula

  The Pythagoras theorem definition shows the relation among the three sides of a right triangle. The square of the hypotenuse is equal to the sum of the square of the other two sides.

  $(\text{Hypotenuse})^2 = (\text{Perpendicular})^2 + (\text{Base})^2$

- Area of a right triangle formula

  The formula to calculate the area of a right triangle formula is given as:

  Area = $1/2 \times \text{Base} \times \text{Height} = 1/2 \times b \times h$

where height,h is equal to the length of the perpendicular side of the triangle.

- The perimeter of a right triangle formula

  The formula to calculate the area of a right triangle formula is given as:

  Perimeter = a + b + c

  where a, b, and c are the three sides of the triangle.

Breakdown tough concepts through simple visuals.

Math will no longer be a tough subject, especially when you understand the concepts through visualizations.

Book a Free Trial Class

Examples Using Right Triangle Formulas

Example 1: The length of the base and perpendicular of a right-angled triangle is 6 in and 8 in respectively. Find:

- Length of its hypotenuse
- The perimeter of the triangle
- Area of the triangle

  Solution:

  To find:

  Given: length of base = 6 in, length of perpendicular = 8 in

  i) Using Pythogoras' theorem,

  $(\text{Hypotenuse})^2 = (\text{Base})^2 + (\text{Perpendicular})^2$

  $(\text{Hypotenuse})^2 = 6^2 + 8^2 = 100$

  Hypotenuse = $\angle 100 = 10$ in

  ii) Using the perimeter of a right triangle formula,

  Perimeter = Sum of all sides

  Perimeter = 6 + 8 + 10 = 24 in

  iii) Using the area of triangle formula,

  Area = $(1/2) \times b \times h$

  $= (1/2) \times 6 \times 8$

  $= 24\ \text{in}^2$

Answer: Hypotenuse of the right triangle = 10 in, the perimeter of the right triangle = 24 in, and the area of the right triangle = 24 in^2.

Example 2: The height and hypotenuse of a right-angled triangle measure 12 in and 13 in respectively. Find its area.

Solution:

To find: Area of a right-angled triangle

Given: Height = 12 in, Hypotenuse = 13 in

Using Pythagoras' theorem,

$(13)^2 = (\text{Base})^2 + (12)^2$

$(\text{Base})^2 = (13)^2 - (12)^2 = 25$

Base = ∠25 = 5 in

Using the Area of a triangle formula,

Area = $(1/2) \times b \times h$

Area = $(1/2) \times 5 \times 12$

Area = 30 in^2

Answer: Area of the right-angled triangle = 30 in^2

Example 3: Determine the area of a right-angled triangle whose perimeter is 30 units, height is 12 units, and the hypotenuse is 13 units

Solution:

To find: Area of a right-angled triangle

Given: perimeter = 30 units, hypotenuse = 13 units, height = 12 units

We know that perimeter = base + hypotenuse + height

30 units = 13 + 12 + base

Therefore, base = 30 - 25 = 5 units

Area = $1/2bh = 1/2(5 \times 12)$ = 30 sq units.

Answer: Area of the right-angled triangle = 30 $unit^2$

10

Radian

Radian is a unit used to measure angles. We have two units to measure the angles: degree and radian. Up to this stage, you might have been using degrees to measure the sizes of angles. However, for a variety of reasons, angle measures in advanced mathematics are frequently described using a unit system different from the degree system. This system is known as the radian system. Did you know that radian was the SI supplementary unit for measuring angles before 1995? It was later changed to a derived unit.

Come, let us learn in detail about the radian formula, the arc length formula, and how to convert an angle from radians to degrees and from degrees to radians.

WHAT IS RADIAN?

The radian is an S.I. unit that is used to measure angles and one radian is the angle made at the center of a circle by an arc whose length is equal to the radius of the circle. A single radian which is shown just below is approximately equal to 57.296 degrees. We use radians in place of degrees when we want to calculate the angle in terms of radius. As '°' is used to represent degree, rad or ^{c} is used to represent radians. For example, 1.5 radians is written as 1.5 rad or 1.5^c.

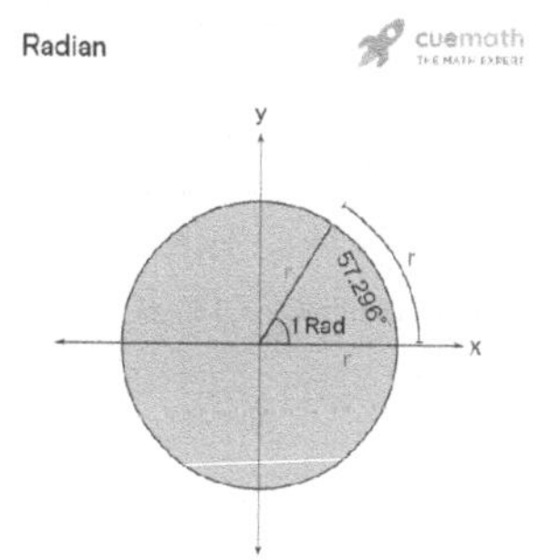

RADIAN DEFINITION

"Radian" is a unit of measurement of an angle. Here are few facts about "radian"

- Radian is denoted by "rad or using the symbol "c" in the exponent.
- If an angle is written without any units, then it means that it is in radians.
- Some examples of angles in radians are, 2 rad, $\pi/2$, $\pi/3$, 6^c, etc.

Radian Uses

- Angles are most or generally measured in radians in calculus and in most other branches of mathematics.
- Radians are widely used in physics also. They are preferred over degrees when angular measurements are done in physics.

Radian Formula

We have already learned that 1 radian is equal to the angle made by the arc of a circle whose length is same as the radius of the circle. Thus, the angle subtended by an arc in radians of a circle is defined as the ratio of the arc length to the radius of the circle.

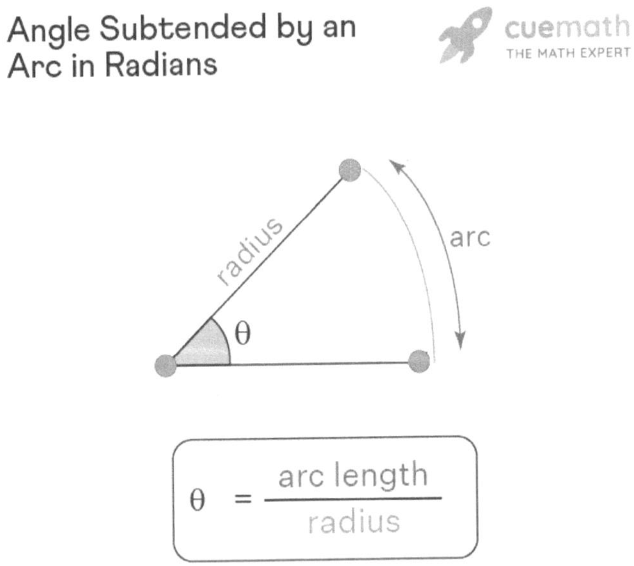

If we consider the arc to be the total circumference of the circle, then arc length = 2Πr. Also, we know that the angle subtended at the center of the circle by its circumference is 360°. Then by the above formula,

Angle subtended = (arc length)/(radius)

$360° = (2\Pi r)/r$

$360° = 2\Pi$

Thus, the formula of radians is 2Π = 360°.

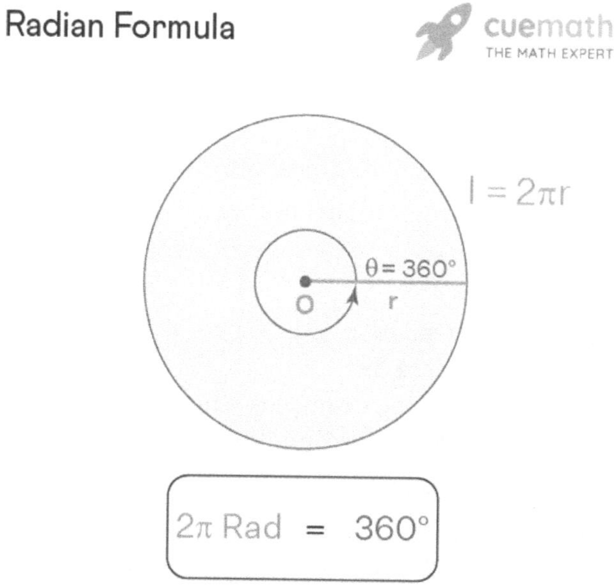

Conversion Between Radians and Degrees

An angle can be converted from "radians to degrees" and from "degrees to radians" according to necessity. We use the radian formula (from the previous section), 2Π = 360° for doing these conversions. We can see how to do the conversions between the radians and degrees in the figure below.

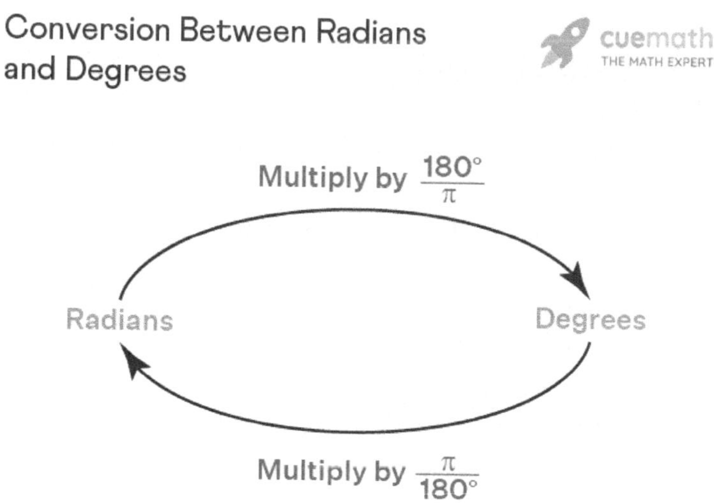

Converting Radians to Degrees

The radian formula can be written as,

2 Radians = 360°

From this, 1 Radian = 360°/2Π (or)

1 Radian = 180°/Π.

Thus, to convert radians to degrees, we multiply the angle by 180°/Π.

Examples of Converting Radians to Degrees:

- $\pi/2 = \pi/2 \times 180°/\pi = 90°$
- $\pi/4 = \pi/4 \times 180°/\pi = 45°$
- $7\pi/6 = 7\pi/6 \times 180°/\pi = 210°$
- 2 rad = $2 \times 180°/\pi$ H$\approx$ 114.59°

If we observe the first three examples where the angle is in terms of Π, Π is getting canceled while converting it into degrees.

So to convert an angle in radians that is in terms of Π into degrees, just replace Π with 180°.

This is a trick to convert radians into degrees. Here you can see the first three examples using the trick.

- $\pi/2 = 180°/2 = 90°$
- $\pi/4 = 180°/4 = 45°$
- $7\pi/6 = 7(180°)/6 = 210°$

Converting Degrees to Radians

From the radian formula,

2Π Radians = 360°

From this, 1° = (2Π Radians)/360°

1° = (Π/180) radians

Thus, to convert degrees to radians, we multiply the angle by Π/180 radians.

Examples of Converting Degrees to Radians:

- $90° = 90 \times \pi/180 = \pi/2$
- $180° = 180 \times \pi/180 = \pi$
- $210° = 210 \times \pi/180 = 7\pi/6$

Radians and Degrees Table

Here is a table with some standard angles in degrees and the corresponding angles in radians. This table is helpful to know the equivalent angles of radians (or degrees).

Degree	Radian
30°	π/6
45°	π/4
60°	π/3
90°	π/2
180°	π
270°	3π/2
360°	2π

Differences Between Radians and Degrees

Radians and degrees both are measurements of angles only. Here are a few differences between radians and degrees.

Radian	Degree
The angle subtended by an arc of length 'r' of a circle whose radius is 'r' is known as 1 radian.	1/360 th part of a complete angle is called a degree.
1 radian is denoted by 1^c or sometimes simply 1.	1 degree is denoted by 1°.
To convert an angle from degrees to radians, multiply it by π/180.	To convert an angle from radians to degrees, multiply it by 180/π.

TRIGONOMETRIC RATIOS IN RADIANS

Introduction

When you delve deeper into your study of trigonometry, you will find that when talking about angle measures, the radian scale is much more widespread than the degree scale. Given this, it is imperative to know the important T-ratio values in the radian scale. These are summarized below (both degree and radian scales are used; when the angle value has no unit specified, it is assumed to be in radians; N.D. means not-defined).

Make sure you read and commit to memory each of these values of trigonometric ratios.

Specific Degree measure in Radians:

1. θ=00orθ=0θ=00orθ=0

- sin00=0orsin0=0sina 00=0orsina 0=0
- cos00=1orcos0=1cosa 00=1orcosa 0=1
- tan00=0ortan0=0tana 00=0ortana 0=0
- cosec00:N.D.orcosec0:N.D.cosec00:N.D.orcosec0:N.D.
- sec00=1orsec0=1seca 00=1orseca 0=1
- cot00:N.D.orcot0:N.D.cota 00:N.D.orcota 0:N.D.

2. θ=300orθ=π6θ=300orθ=π6

- sin300=12orsinπ6=12sina 300=12orsina π6=12
- cos300=∠32orcosπ6=∠32cosa 300=32orcosa π6=32
- tan300=1∠3ortanπ6=1∠3tana 300=13ortana π6=13
- cosec300=2orcosecπ6=2cosec300=2orcosecπ6=2
- sec300=2∠3orsecπ6=2∠3seca 300=23orseca π6=23
- cot300=∠3orcotπ6=∠3cota 300=3orcota π6=3

3. θ=450orθ=π4θ=450orθ=π4

- sin450=1∠2orsinπ4=1∠2sina 450=12orsina π4=12
- cos450=1∠2orcosπ4=1∠2cosa 450=12orcosa π4=12
- tan450=1ortanπ4=1tana 450=1ortana π4=1
- cosec450=∠2orcosecπ4=∠2cosec450=2orcosecπ4=2
- sec450=∠2orsecπ4=∠2seca 450=2orseca π4=2
- cot450=1orcotπ4=1cota 450=1orcota π4=1

4. θ=600orθ=π3θ=600orθ=π3

- sin600=∠32orsinπ3=∠32sina 600=32orsina π3=32
- cos600=12orcosπ3=12cosa 600=12orcosa π3=12
- tan600=∠3ortanπ3=∠3tana 600=3ortana π3=3
- cosec600=2∠3orcosecπ3=2∠3cosec600=23orcosecπ3=23
- sec600=2orsecπ3=2seca 600=2orseca π3=2
- cot600=1∠3orcotπ3=1∠3cota 600=13orcota π3=13

5. θ=900orθ=π2θ=900orθ=π2

- sin900=1orsinπ2=1sina 900=1orsina π2=1
- cos900=0orcosπ2=0cosa 900=0orcosa π2=0
- tan900:ndortanπ2:ndtana 900:ndortana π2:nd
- cosec900=1orcosecπ2=1cosec900=1orcosecπ2=1
- sec900:ndorsecπ2:ndseca 900:ndorseca π2:nd
- cot900=0orcotπ2=0

11

Heights and Distances

Trigonometry is the study of the relationship between the length of sides and angles of a triangle. A triangle is a closed shape consisting of three sides. The relation between the heights and distances of objects can be understood using trigonometry.

Trigonometry is one of the most ancient subjects studied by scholars all over the world. Trigonometry was invented because its need arose in astronomy. Since then astronomers have used it, for instance, to calculate distances from the Earth to the planets and stars. Trigonometry is also used in geography and in navigation.

How Tall is the Glacier ?

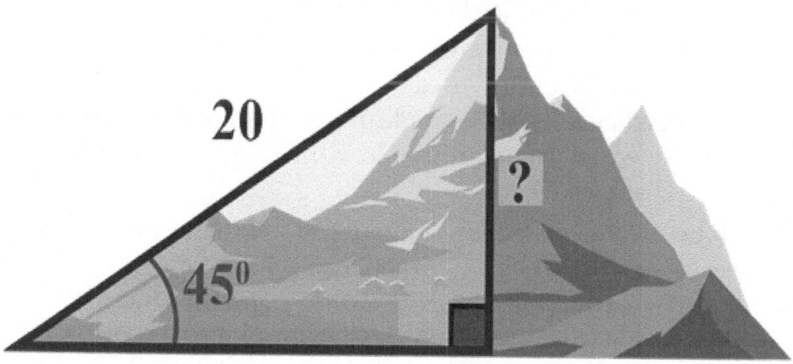

The knowledge of trigonometry is used to find heights of structures, construct maps, determine the position of an island in relation to the longitudes and latitudes.

Trigonometry has found its importance in multiple fields. It has found its applications in the fields ranging from Engineering to Architecture to Astronomy as well. Trigonometry can be used in these fields by to measure distances and angles by assuming lines that connect the points.

Height is the measurement of an object in the vertical direction and distance is the measurement of an object from a particular point in the horizontal direction. If we imagine a line connecting the point of observation to the topmost point of the object then the horizontal line, vertical line and the imaginary line will form a triangle.

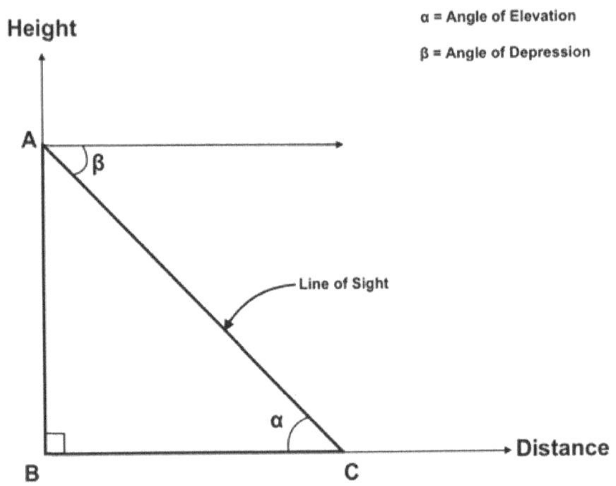

Observe the figure. Consider the observer to be at point C. The height of the object is shown by line AB. The distance of the object from the observer is given by line BC. The object may or may not be perpendicular to the ground. Line AC represents the Line of Sight when the observer is observing the topmost point of the object. Angle α represents the angle of elevation and Angle â represents the angle of depression.

Using trigonometry, if we are provided with any of the two quantities that may be a side or an angle, we can calculate all the rest of the quantities. By the law of alternate angles, the angle of elevation and angle of depression are consequently equal in magnitude ($\alpha = \beta$). Tan α is equal to the ratio of the height and distance.

You can be provided with any two of the following information:

- The distance of the object from the observer
- The height of the object
- Angle at which the observer views the topmost point of the object (angle of elevation)
- The angle at which the observer views the object when the observer is on top of a tower/building (angle of depression)

and you can calculate the rest. Here are some examples for you in the following section.

The most significant definitions that are used when dealing with heights and distances are given as:

Line of sight: It is the line drawn from the eye of an observer to the point in the object viewed by the observer.

Here, the cat is the observer and the object is the bird.

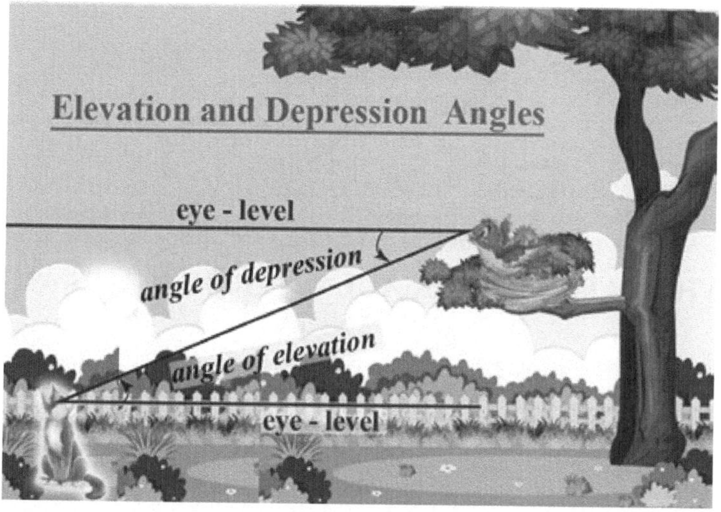

The angle of elevation: The angle between the horizontal and the line of sight joining an observation point to an elevated object.

In the following figure, the angle of elevation of the kite from the point AA is 30°30°, and from point BB is 60° 60°

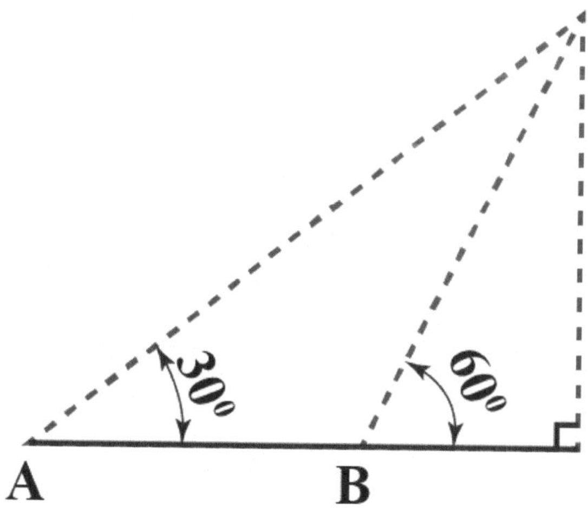

The angle of depression: The angle between the horizontal and the line of sight joining an observation point to an object below the horizontal level.

In the following figure, if the top of the building is our observation point, then the angle of depression of person XX is 45 45′″, and that of person YY is 60′″60′″.

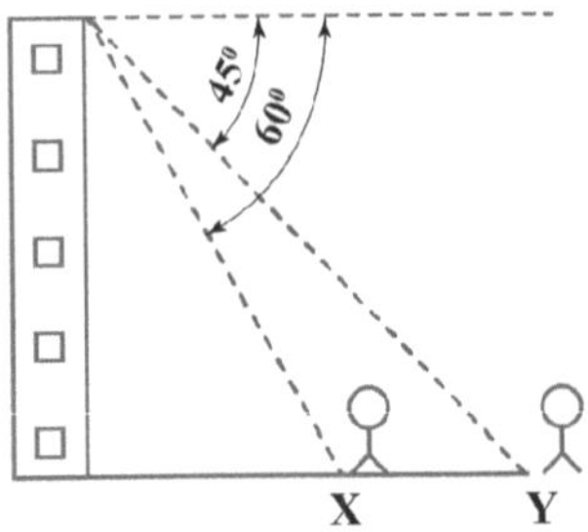

Let us now experience how trigonometry is applied to practical situations.

WHAT ARE THE FORMULAS FOR HEIGHTS AND DISTANCES?

Trigonometric ratios can be used to find heights and distances. Some useful relations are illustrated by the below-given diagrams which help us to determine heights and distances.

In a triangle ABC,

$$\sin\theta = \frac{h}{l}, \cos\theta = \frac{b}{l} \text{ an } \tan\theta = \frac{h}{b}$$

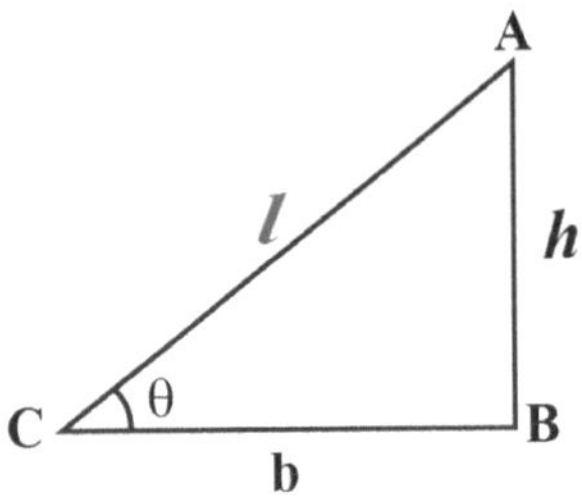

In a triangle ABC,

if ED || AB then, $\frac{AB}{ED} = \frac{BC}{DC}$

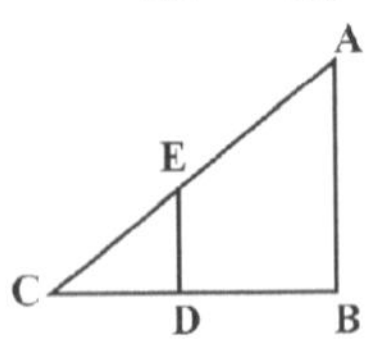

$d = h(\cot\alpha - \cot\beta)$

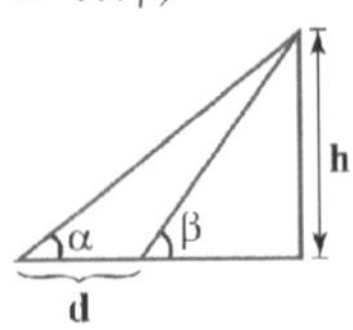

1. The angles of depression of the top and the bottom of an 8yd8yd tall building from the top of a multi-storeyed building are 30°30° and 45°45°, respectively. Find the height of the multi-storeyed building and the distance between the two buildings.

Hint: Try to depict the figure according to a given situation.

How to Find Height and Distances?

To measure the heights and distances of different objects, we use trigonometric ratios.

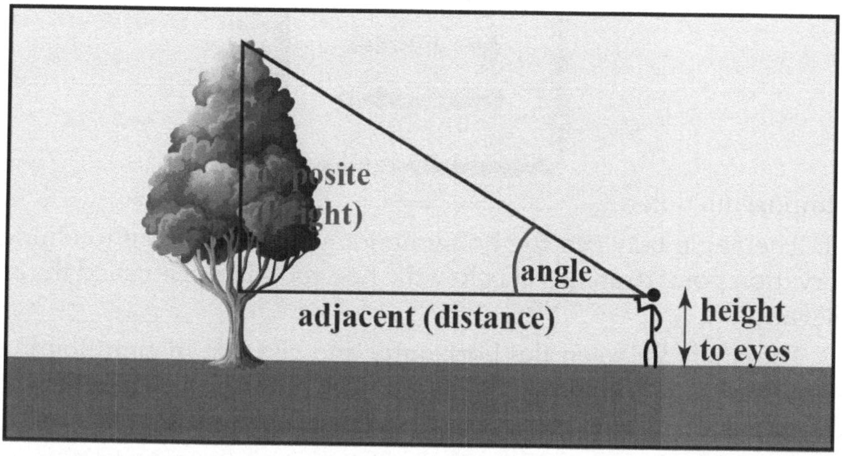

Use the Tangent rule to calculate the height of the tree (above eye level).

tan(angle) = opposite/adjacent

Where the opposite is the height of the tree and adjacent is the distance between you and the tree.

This is rearranged to:

opposite = tan(angle) x adjacent

or more simply

height=tan(angle)×distanceheight=tan(angle)×distance

Distance can be calculated as:

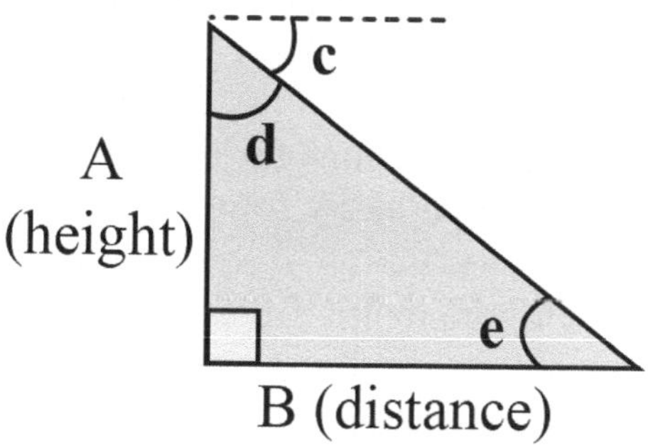

B (distance)=A (height)tan (e)B (distance)=A (height)tan (e)

Therefore, to calculate BB (distance) we will need the value of AA (height) and angle ee.

Important Notes

1. The angle between the horizontal and the line of sight joining an observation point to an object below the horizontal level is called the angle of elevation.

2. The angle between the horizontal and the line of sight joining an observation point to an object below the horizontal level is called the angle of depression.

3. B (distance)=A (height)tan (e)B (distance)=A (height)tan (e)

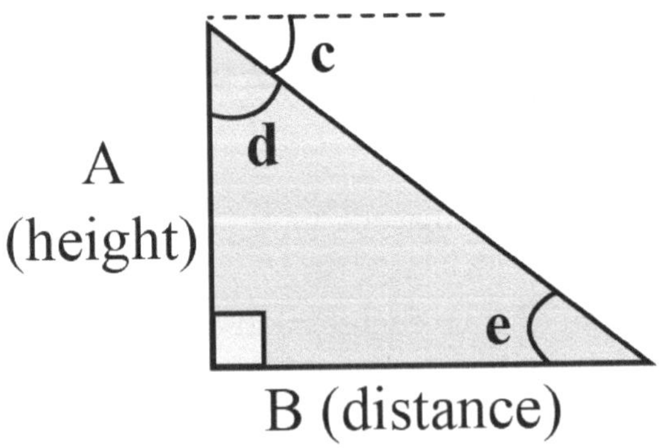

Trigonometric Ratios Table

Values of trigonometric functions for the standard angles such as 0°, 30°, 45°, 60°, and 90° could be easily found using a trigonometric ratios table.

The table consists of trigonometric ratios – sine, cosine, tangent, cosecant, secant, and cotangent.

In short, these ratios are written as sin, cos, tan, cosec, sec, and cot.

It is best to remember the values of the trigonometric ratios of these standard angles.

degrees	radians	sin θ	cos θ	tan θ	csc θ	sec θ	cot θ
0^0	0	0	1	0	Not defined	1	Not defined
30^0	$\frac{\pi}{6}$	$\frac{1}{2}$	$\frac{\sqrt{3}}{2}$	$\frac{\sqrt{3}}{3}$	2	$\frac{2\sqrt{3}}{3}$	$\sqrt{3}$
45^0	$\frac{\pi}{4}$	$\frac{\sqrt{2}}{2}$	$\frac{\sqrt{2}}{2}$	1	$\sqrt{2}$	$\sqrt{2}$	1
60^0	$\frac{\pi}{3}$	$\frac{\sqrt{3}}{2}$	$\frac{1}{2}$	$\sqrt{3}$	$\frac{2\sqrt{3}}{3}$	2	$\frac{\sqrt{3}}{3}$
90^0	$\frac{\pi}{2}$	1	0	Not defined	1	Not defined	0

A trigonometric table has wide application in fields like science and engineering.

Solved Examples

Example 1

A man standing at a certain distance from a building, observe the angle of elevation of its top to be 60′″60′″. He walks 30 yds away from the building. Now, the angle of elevation of the building's top is 30′″30′″. How high is the building?

Solution

Let the height of the building be hh, and dd be the original distance between the man and the building. The following figure depicts the given situation:

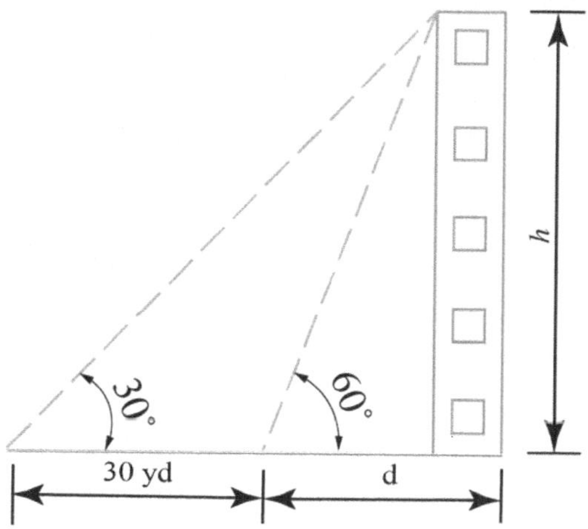

We have:

tan60°=√3⇒hd=√3⇒d=h√3tana 60°=3⇒hd=3⇒d=h3

Also,

tan30°=1√3⇒hd+30=1√3⇒√3h=d+30⇒√3h=h√3+30⇒h(√3−1√3)=30tana 30°=13⇒hd+30=13⇒3h=d+30⇒3h=h3+30⇒h(3−13)=30

⇒h=15√3yd≈26yd⇒h=153ydH≈26yd

∴ ∴ The building's height is about 26 yards.

Example 2

From a certain point on the ground, the angle of elevation of the top of a tree is αα. On moving pp meters towards the tree, the angle of elevation becomes ββ . Show that the height of the tower is

h=(ptanαtanβtanβ-tanα)metresh=(ptana αtana βtana βtana α)metres

Solution

Observe the following figure, which depicts this situation:

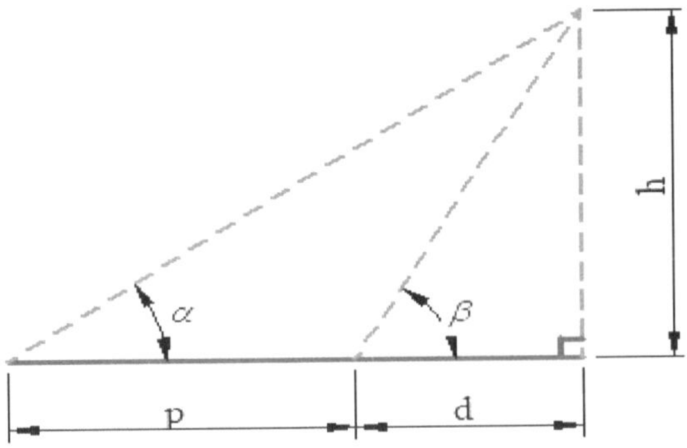

Here, dd and hh are unknown and we need to find hh .We have :

tanβ=hd⇒d=hcotβtanaβ=hd⇒d=hcota β

Now, we have,

tanα=hp+d⇒h=(p+d)tanα=(p+hcotβ)tanα=ptanα+htanα cotβ⇒h(1−tanαcotβ)=ptanα⇒h=ptanα1−tanαcotβ=ptanα1−tanαtanβtana α=hp+d⇒h=(p+d)tana α=(p+hcotaβ)tana α=ptana α+htana αcota β⇒h(1−tana αcotaβ)=ptana α⇒h=ptana α1−tana αcota β=ptana α1−tana αtanaβ

h=ptanαtanβtanβtanαh=ptana αtana βtana βtana α

∴ ∴h=ptanαtanβtanβtanαh=ptana αtana βtana βtana α

Example 3

A house has a window hh yards above the ground. Across the street from this house, there is a tall pole. The angle of elevation and depression of the top and bottom of this pole from the window are θ and φφ respectively. Determine the height of the pole.

Solution

The following figure depicts the given situation:

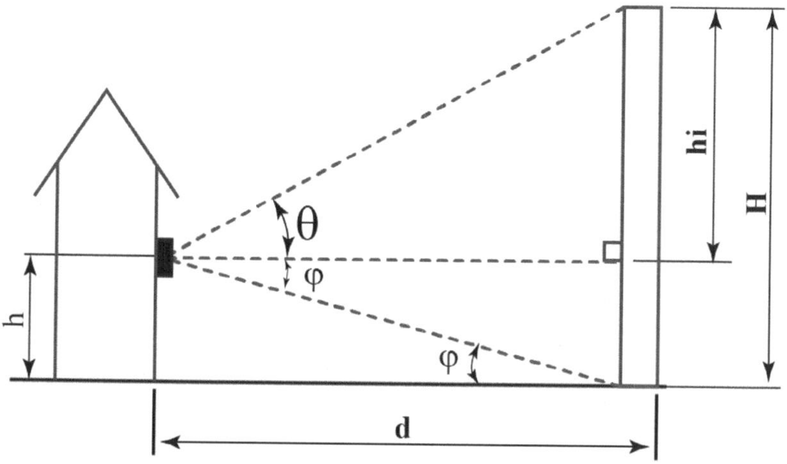

Note that d=hcotφd=hcota φ and

h1=dtanθ=htanθcotφh1=dtana θ=htana θcota φ

Thus, the height of the pole is,

H=h+h1⇒H=h(1+tanθcotφ)H=h+h1⇒H=h(1+tana θcota φ)

∴ ∴ The height of the pole is =h(1+tanθcotφ)=h(1+tana θcota φ)

Example 4

From an observation tower, the angle of depression of two cars on the opposite side of the tower are ααand ββ. If the tower's height is hh yards, find the distance between the cars.

Solution

Observe the following figure:

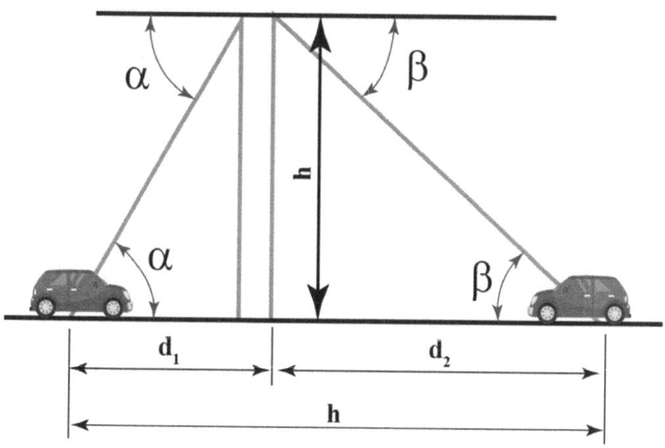

We have,

$d1=h\cot\alpha,d2=h\cot\beta d1=h\mathrm{cota}\ \alpha,d2=h\mathrm{cota}\ \beta$

Thus, the distance between the cars is,

$D=d1+d2=h(\cot\alpha+\cot\beta)ydD=d1+d2=h(\mathrm{cota}\ \alpha+\mathrm{cota}\ \beta)yd$

$\therefore\therefore$ The distance between the cars is$=h(\cot\alpha+\cot\beta)yd=h(\mathrm{cota}\ \alpha+\mathrm{cota}\beta)yd$

Bibliography

- Abeles, Francine F. (2008). Dodgson condensation: The historical and mathematical development of an experimental method. Linear Algebra and its Applications. 429 (2–3): 429–438. doi:10.1016/j.laa.2007.11.022.
- Anton, Howard (2005), Elementary Linear Algebra (Applications Version) (9th ed.), Wiley International
- Axler, Sheldon Jay (2015). Linear Algebra Done Right (3rd ed.). Springer. ISBN 978-3-319-11079-0.
- Bareiss, Erwin (1968), Sylvester's Identity and Multistep Integer-Preserving Gaussian Elimination (PDF), Mathematics of Computation, 22 (102): 565–578, doi:10.2307/2004533, JSTOR 2004533
- Blass, Andreas (1984), Existence of bases implies the axiom of choice (PDF), Axiomatic set theory (Boulder, Colorado, 1983), Contemporary Mathematics, vol. 31, Providence, R.I.: American Mathematical Society, pp. 31–33, MR 0763890
- Bourbaki, Nicolas (1998), Algebra I, Chapters 1-3, Springer, ISBN 9783540642435
- Brown, William A. (1991), Matrices and vector spaces, New York: M. Dekker, ISBN 978-0-8247-8419-5
- Bunch, J. R.; Hopcroft, J. E. (1974). Triangular Factorization and Inversion by Fast Matrix Multiplication. Mathematics of Computation. 28 (125): 231–236. doi:10.1090/S0025-5718-1974-0331751-8.
- Cédric Jules (February 25, 2015). 2D transformation matrices baking.
- de Boor, Carl (1990), An empty exercise (PDF), ACM SIGNUM Newsletter, 25 (2): 3–7, doi:10.1145/122272.122273, S2CID 62780452
- Dummit, David S.; Foote, Richard M. (2004), Abstract algebra (3rd ed.), Hoboken, NJ: Wiley, ISBN 9780471452348, OCLC 248917264
- Fang, Xin Gui; Havas, George (1997). On the worst-case complexity of integer Gaussian elimination (PDF). Proceedings of the 1997 international

symposium on Symbolic and algebraic computation. ISSAC '97. Kihei, Maui, Hawaii, United States: ACM. pp. 28–31. doi:10.1145/258726.258740. ISBN 0-89791-875-4. Archived from the original (PDF) on 2011-08-07. Retrieved 2011-01-22.

- Fisikopoulos & Peñaranda 2016, §1.1, Bareiss 1968
- Fisikopoulos, Vissarion; Peñaranda, Luis (2016), Faster geometric algorithms via dynamic determinant computation, Computational Geometry, 54: 1–16, doi:10.1016/j.comgeo.2015.12.001
- G. Baley Price (1947) Some identities in the theory of determinants, American Mathematical Monthly 54:75–90 MR0019078
- Garibaldi, Skip (2004), The characteristic polynomial and determinant are not ad hoc constructions, American Mathematical Monthly, 111 (9): 761–778, arXiv:math/0203276, doi:10.2307/4145188, JSTOR 4145188, MR 2104048
- Gentle, James E. (2007). Matrix Transformations and Factorizations. Matrix Algebra: Theory, Computations, and Applications in Statistics. Springer. ISBN 9780387708737.
- Habgood, Ken; Arel, Itamar (2012). A condensation-based application of Cramer's rule for solving large-scale linear systems (PDF). Journal of Discrete Algorithms. 10: 98–109. doi:10.1016/j.jda.2011.06.007.
- Halmos, Paul Richard (1974) [1958]. Finite-Dimensional Vector Spaces (2nd ed.). Springer. ISBN 0-387-90093-4.
- Harris, Frank E. (2014), Mathematics for Physical Science and Engineering, Elsevier, ISBN 9780128010495
- Horn, Roger Alan; Johnson, Charles Royal (2018) [1985]. Matrix Analysis (2nd ed.). Cambridge University Press. ISBN 978-0-521-54823-6.
- http://ocw.mit.edu/courses/aeronautics-and-astronautics/16-07-dynamics-fall-2009/lecture-notes/MIT16_07F09_Lec03.pdf[bare URL PDF]
- J. W. P. Hirschfeld (1979) Projective Geometry of Finite Fields, Clarendon Press
- Katznelson, Yitzhak; Katznelson, Yonatan R. (2008). A (Terse) Introduction to Linear Algebra. American Mathematical Society. ISBN 978-0-8218-4419-9.
- Kleiner, Israel (2007), Kleiner, Israel (ed.), A history of abstract algebra, Birkhäuser, doi:10.1007/978-0-8176-4685-1, ISBN 978-0-8176-4684-4, MR 2347309
- Kung, Joseph P.S.; Rota, Gian-Carlo; Yan, Catherine (2009), Combinatorics: The Rota Way, Cambridge University Press, ISBN 9780521883894.

Index

9 789360 843007

Printed at Repro India Ltd.